THE FUTURE OF ANARCHISM

THE FUTURE OF ANARCHISM

Hugh Small

By the same author:

Florence Nightingale, Avenging Angel (Constable, 1998)

Florence Nightingale and Her Real Legacy – a Revolution in Public Health (Robinson/Little, Brown 2017)

The Crimean War: Europe's Conflict with Russia (History Press 2018)

Copyright © Hugh Small 2017

Published by Knowledge Leak
www.knowledgeleak.com

www.hugh-small.co.uk

ISBN: 9780957279759

Release 2.1.2

19 November 2025

Contents

It is now no longer a question of accumulating scientific truths and discoveries. We need above everything to spread the truths already mastered by science, to make them part of our daily life, to render them common property.

Peter Kropotkin: *An Appeal to the Young*, 1886

CHAPTER 1

Anarchism: Fossil or Future?

THE SEASIDE TOWN OF SWANAGE IN SOUTHERN ENGLAND lies at the eastern end of the Jurassic Coast, a 100-mile long World Heritage Site famous for its eroded cliffs that first made fossil-hunting popular in the early 1800s. By 1885 the town had become a fashionable resort accessible through the national railway network. But in 1972 the state-owned railway operator disconnected Swanage from its network, tore up the 11 miles of track, and sold the railway station and the right-of-way to the town council. The plan was to turn the right-of-way into a fast motor road.

Local residents, who had unsuccessfully campaigned for four years to keep the line open, lobbied the town council for three more years during which the railway buildings were vandalised. Finally, the town council held a referendum in which 83 per cent of residents (Swanage has a population of 9,000) voted to lease the station and right-of-way to a residents' committee. Volunteers began restoration. Within ten years they had laid a mile of track, built a new station at the other end of it, bought rolling stock, and were running passenger services. After another decade, the track extended five miles to the tourist attraction of Corfe Castle. After another decade and a half, trains were again running from London to Swanage.

This example shows that humans are capable of cooperation on a grand scale. Not just the 9000 townspeople and volunteers from outside, but also the annual 200,000 tourist passengers who pay to keep the service running. All are involved in a complex cooperative venture, of which more examples will be given in later chapters.

The Swanage railway cooperative kept start-up costs down by using obsolete and labour-intensive steam and diesel locomotives. The labour has been provided free by enthusiasts who grew up in an era when heavy machinery was glamourous. It is becoming harder to find local people who want to learn to drive a steam locomotive, and climate change and air pollution are making the use of fossil fuels less acceptable. But the temporary expedient has served to leapfrog investment in intermediate technologies like overhead power lines. The future lies in hybrid or battery-driven locomotives, which can use the 10,000 miles of existing track throughout the country.

The Swanage Railway, opposed by the state, is therefore well-positioned to become the most modern railway in Britain. Already this cooperative venture, which is 'contrary to authority' and therefore by definition anarchist, has expropriated the proud boast of modern history's most authoritarian governments.

It has made the trains run on time.

Old questions, new answers

Why do humans so naturally join together in complex ventures? Recent scientific discoveries have shed new light on this question. Archaeologists, geographers, and other academics have shown that natural selection created early humans as compulsive cooperators, rivalling the bees and ants.

Early chapters of this book will show that natural selection did not achieve this by moulding our 'instincts' or culture but by shaping our physiology – the structure of our bodies – so that individual humans could not even survive without cooperating. It was natural selection's response to a harsh environment that no longer exists. Cooperative humans as we know them could not evolve in today's climate. Nevertheless, we are here, condemned to be what mathematician

Martin Novak calls 'super-cooperators' even though survival no longer requires it.

The recent discovery of human cooperation has turned the scientific consensus on its head. Until two decades ago humans were thought to have evolved as selfish individualists, only capable of peaceful coexistence when controlled by hierarchical government. If it seems hard to believe that scientists could get it so wrong, remember zoologist Richard Dawkins' famous comment that "we are born selfish" in his 1976 book *The Selfish Gene*. His retraction of this statement thirty years later in the foreword to his 2006 edition of the same book is an indicator of a radical change in the scientific consensus. His recantation is not as widely known to the public as his original theory.

There was originally a grain of truth in the received wisdom that humans needed to be controlled by hierarchical government. That style of government was necessary when the first states were formed by conquest and exploitation of one tribe by another. As conquest gradually went out of use and populations became less rebellious, it escaped notice that the only remaining beneficiaries of hierarchical government were those at the top of the hierarchy.

Defenders of the hierarchical *status quo* have denied the importance of the new scientific discoveries. While accepting that it invalidates the long-held justification for government, they are using new evidence-free hypotheses to explain why we should retain a political system that was originally developed to subdue conquered populations. For example Yuval Noah Harari, the author of *Sapiens*, implicitly recognises the new discoveries about human cooperation and agrees that government was not always needed. But he maintains that innate human cooperation could not cope with the spread of agriculture 12,000 years ago and subsequent city-building:

> The problems at the root of such calamities [e.g. wars] is that humans evolved for millions of years in small bands of a few dozen individuals. The handful of millennia separating the Agricultural Revolution from the appearance of cities, kingdoms and empires was not enough time to allow an

instinct for mass cooperation to evolve … Unfortunately complex human societies seem to require imagined hierarchies and unjust discrimination.[1]

Harari's assertion is that natural selection evolved 'instincts' for human cooperation that worked for Stone Age hunters and gatherers but do not work in the modern population. According to Harari, the solution to such evolutionary shortcomings is the necessary evil of hierarchical government. According to him the Swanage railway was impossible, involving as it did thousands of townspeople and hundreds of thousands of visitors, who together corrected the errors of the hierarchical state.

If there are any readers who feel reassured by popular explanation of why human societies still need hierarchical government, and why we should ignore changes in the scientific consensus, the rest of this book will probably make them uncomfortable.

Harari's version of evolution uses the outdated term 'instinct' to attribute human cooperative behaviour to some mysterious fragment of the brain. On the contrary, our dependence on cooperation arose because natural selection eliminated body structures that cause other primates to compete against members of the same species.

This book will test Harari's assertion that hierarchical government corrects human evolutionary limitations. It will do so by examining a recent scientific discovery about speech communications. Social psychologists have found that our communication system, which evolved because it allowed humans to pass information to the next generation, cannot reliably distinguish truth from falsehood without careful training. If Harari were right that hierarchical government is needed to solve our evolutionary shortcomings, it should have provided for this training. On the contrary this book will show that today's hierarchical governments have made the problem worse.

Other popular science books have not placed our newly discovered cooperative nature in a social context. This book does so. It attempts "to spread the truths already mastered by science, to make them part of our daily life, to render them common property" as the anarchist and scientist Peter Kropotkin put it. It summarises the last two decades' discoveries about the origin of human society and reinterprets history in the light of the new scientific knowledge. Completely anarchist (i.e. non-hierarchical) society will emerge in accordance with this new knowledge. Today's democratic hierarchical government was conceived in the light of scientific opinion which is now known to be outdated.

Chapter 2 shows that Kropotkin's view of how natural selection created human cooperation was more scientific than that of his contemporary, Darwin. Darwin and his followers believed that competition between humans caused our evolutionary divergence from other primates. This off-the-top-of-the head view of human society, quite separate from Darwin's carefully researched theory of evolution, created a pseudo-scientific justification for the perpetuation of the hierarchical state.

Chapter 3 traces the origins and limitations of human speech communication and shows why our culture is vulnerable to propaganda and thought control when social hierarchies arise.

Chapters 4 through 7 describes how global environmental changes enabled social hierarchies and states to come into existence. This terminated a long, formative period of egalitarian anarchism in human society. I will show how a hierarchy automatically imposes thought control which ensures its own preservation.

Chapter 8 describes the 200-year experiment in parliamentary electoral democracy. I demonstrate how modern Britain has realised the worst fears of democracy's founders.

The concluding chapter shows how to convert a modern state into a democratic cooperative. It will describe how 'anarchist democracy' can use the electoral process to install a non-hierarchical executive. This will come about when a critical mass of the population has understood the science behind cooperation and recognised the obsolescence and ineffectiveness of our present system of government.

CHAPTER 2

Why Humans Cooperate

CHARLES DARWIN BELIEVED THAT A 'struggle for existence' enabled cooperative traits to survive by natural selection. In the case of humans, he believed that the struggle was against other humans, in warfare. Unaware of the very limited population of humans during our evolutionary period, and their wide dispersal, Darwin imagined that it was through never-ending warfare that 'the social and moral qualities' – meaning cooperation and loyalty – evolved to become innate, as he wrote in *The Descent of Man*:

> When two tribes of primeval man, living in the same country, came into competition, if (other circumstances being equal) the one tribe included a great number of courageous, sympathetic and faithful members, who were always ready to warn each other of danger, to aid and defend each other, this tribe would succeed better and conquer the other. Let it be borne in mind how all-important in the never-ceasing wars of savages, fidelity and courage must be. ... A tribe rich in the above qualities would spread and be victorious over other tribes: but in the course of time it would, judging from all past history, be in its turn overcome by some other tribe still more highly endowed.

Thus the social and moral qualities would tend slowly to advance and be diffused throughout the world.

In arriving at this conclusion Darwin was influenced by the claims of Thomas Robert Malthus that population would always outpace food production and that this would lead to inevitable culling of the population. Darwin noted in his *Autobiography*:

> I happened to read for amusement Malthus on *Population*, and being well prepared to appreciate the struggle for existence … it at once struck me that under these circumstances favourable variations would tend to be preserved and, and unfavourable ones to be destroyed. The result of this would be the formation of new species. Here then I had at last got a theory by which to work.'[2]

Speculating well outside his area of expertise, it is not surprising that Darwin should think that the wars of savages had always been 'never-ceasing'. Wherever Europeans had first encountered native tribes, as they were doing constantly during Darwin's lifetime, they carried diseases to which they had become immune through millennia of close contact with farm animals. Indigenous people fled before them and created a domino effect, clashing with other tribes over safe living space. The conflicts were fuelled by European firearms and horses. Throughout the first decades of Darwin's life, the Maori tribes of New Zealand massacred each other in what became known as the 'Musket Wars'. In Africa, Shaka Zulu drove inter-tribal war to new heights of cruelty, and in America the native tribes poured onto the plains from the eastern settlements that were being invaded by Europeans and began to fight each other for survival. Newly-arrived Europeans reported these conflicts as the customary way of life of 'savages'.

Darwin knew nothing about the diseases carried by Europeans and their effectiveness in driving native tribes to extinction, very little of the human archaeological record or early human demography, and nothing about the history of 'savages' which would have revealed that what he called their 'never-ceasing wars' were a phenomenon which (it is now known) came too late to have influenced the evolution of innate cooperative behaviour.

Warfare or mutual aid?

The naturalist Peter Kropotkin (1842-1921) promoted Darwin's theory of evolution, and his analysis of the struggle for survival took account of information that was not available to Darwin or to Malthus. Kropotkin was a Russian prince, a member of the dynasty which had ruled Russia before the Romanov Tsars. He had graduated top of the class in the most exclusive military school in St. Petersburg and had become the personal page of Tsar Alexander II. The reforms of Catherine the Great had encouraged aristocrats to move from the court to their country estates whose slave labourers, the serfs, were the source of their wealth. Until then, the urban aristocracy had regarded the Russian hinterland as a colony inhabited by an inferior race. Once the aristocracy moved to the country their children, like Peter Kropotkin, engaged with the vast mass of the Russian population and began to harbour ideas of liberating and improving them.

Kropotkin's early career was as a military administrator in Siberia during the rule of the Tsars. Although backward by European standards, Russia had an efficient system for autocratic control of its remote settlements and prison camps. Its telegraph network was extensive, and a system of government post-houses enabled Kropotkin to travel thousands of miles on official business by sleigh and carriage at an average of ten miles per hour day and night. His experiences led him to abandon his military career and to study at St. Petersburg University. He became secretary of the Russian Geographic Society and published many papers on the geography and prehistoric climate of Russia. His political activism led to imprisonment, but he escaped and travelled to Western Europe.

Most European states were happy to extradite Russian dissidents so that their native country could entomb them in fortress prisons or exile them to the gulag. Kropotkin's brother Alexander suffered the latter fate and killed himself. Peter Kropotkin was imprisoned in

France but after some years of public agitation in his favour he was released and went into exile in England.

Kropotkin's extensive travel in Eurasia had led him to conclude that Malthus's argument was entirely based on the special circumstances of England, one of the most overcrowded and underfed countries in the world. Kropotkin had found no trace in the Eurasian land mass of starvation or conflict over resources. Animal populations were limited not by food supply but by extremes of weather. Malthus's original inspiration had come from the parish registers of his small and leafy parish in Suffolk. He had noticed that births were outnumbering deaths and the population was growing unhealthy and stunted because food production was not keeping pace. It is not likely that the register of a small parish on the steppe would have shown the same pattern.

Kropotkin agreed enthusiastically with Darwin's theory of natural selection but argued that Darwin's information-gathering voyage on the *Beagle* had biased his sample of wildlife towards competitive species.[3] Darwin had mainly visited equatorial coastal zones, where overcrowding was more common and climate more benign than in the higher latitude regions studied by Kropotkin. Darwin did not study sparse animal populations, extreme seasonal weather variations, or continental ecology of the type familiar to Kropotkin.

Kropotkin was influenced by Karl Fedorovich Kessler, Head of the Zoology Department at St. Petersburg University. Kessler, while supporting Darwin's theory of natural selection, believed that Darwin had overrated the importance of competition between members of the same species. He identified competition's *opposite* – what he called 'mutual aid' – as another behaviour that could evolve by natural selection. Kropotkin, based on his own observations in Siberia, expanded Kessler's argument by proposing that the 'struggle for existence' could be against a harsh climate, not just against other members of the same species.

In taking this approach, Kropotkin is falsely accused of believing in humanity's innate 'virtue' and even of rejecting the concept of evolution by natural selection. A typical dismissal is by zoologist Matt

Ridley, in *The Origins of Virtue*: 'Kropotkin's was not a mechanistic theory of evolution, like Darwin's'.

The continued misrepresentation of Kropotkin's ideas in this way has encouraged the view that political anarchism is for sentimentalists who believe in the perfectibility of humankind (an idea that critics discover under every liberal's bed). Kropotkin had no such illusions. For him, anarchism was a policy that makes maximum use of mutual aid as an alternative to unnatural and economically destructive hierarchical control.

Many modern biologists are preoccupied, as Darwin was, with explaining how natural selection could occasionally appear to evolve unselfish ('altruistic') behaviour. Common examples of animals that appear to behave altruistically are species like bees and ants of which some individuals give up their right to reproduce. Historians of science have assumed that Kropotkin was on the same quest for altruism, and he has been criticised for thinking that 'mutual aid' between members of the same species was altruism under another name. Lee Alan Dugatkin, in his book *The Altruism Equation* wrote:

> For Kropotkin, group life *per se*, and indeed almost every sort of action involving members of the same species — with the exception of aggression, which he hardly ever recorded — constituted altruism.[4]

This is not true. Kropotkin was not interested in 'altruism' at all. In his book *Mutual Aid: A Factor in Evolution* he did not use the term, and went out of his way to politely ridicule the leading scientists of his time who tried to 'prove the existence of love and sympathy among animals'.[5]

What Kropotkin wanted to do in *Mutual Aid* was to undercut Darwin's claim that natural selection used *competition* between members of one species to ensure that only the fittest would survive. On the contrary, Kropotkin's evidence showed that natural selection more often favours traits that *reduce* competition. The type of behaviour he studied was that in which a species evolves to act in harmony in a way that benefits all equally. Creating quadrupeds that herd together, or birds that form flocks, or fish that shoal is a more effective way for natural selection to ensure individual survival than evolving more armour or better individual evasion skills. It is, as he was at pains to point out, not self-sacrifice because all benefit equally.

Kropotkin pointed out that Darwin gave no examples of direct competition between members of the same species. Darwin's omission appears strange when competition for access to mates is so common. Kropotkin's explanation for the omission is that Darwin, like Malthus, was obsessed with the problem of overpopulation. 'He [Darwin] often speaks of regions being stocked with animal life to their full capacity, and from that overstocking he infers the necessity of competition.'[6] Competition for mates does nothing to reduce population, so it was not relevant to Darwin's theory and he ignored it.

Kropotkin saw a struggle against extreme weather as something that natural selection would find a way to avoid through mutual aid. He cited mass seasonal migration as one way that natural selection can adapt a species to extreme weather without making them struggle against it. The chance mutations needed to establish such behaviour in each individual are few and minor: a tendency to stay together with other members of the species and a sense of direction.

Kropotkin wrote *Mutual Aid* late in life and had studied all the available literature to back up his own observations in Eurasia. The vast number of 'sociable' species that he documented in detail showed that nature often creates both cooperation and competition. A species may demonstrate both – as when hoofed animals herd together for protection but males still fight for access to females. Mutual aid may be more common in nature than either competitive or altruistic behaviour.

Kropotkin went on to show that modern humans, too, are very effective at mutual aid. He showed that throughout history federations of cooperating local groups created more effective government than centralised hierarchies. This conclusion was reached independently by his contemporary, the academic sociologist Max Weber.

The timeline of human cooperation

Darwin's guess that human cooperation evolved through warfare was not based on evidence. Kropotkin's theory that it evolved in response to climate was supported by observation of animal species in harsh climates. There was virtually no human fossil evidence in their time. To decide who was nearer the truth, Darwin or Kropotkin, we can examine what is now known about the climate in which humans evolved. We can align climate change with the development of human cooperation as seen in the fossil record.

Until the beginning of the 21st century anthropologists believed that early humans had a relatively easy life in the Stone Age. Scientists relied on the 'savannah hypothesis' to describe the environment in which humanity evolved. According to this hypothesis, developed by Raymond Dart shortly after Kropotkin's death, the first humans evolved to suit Africa's wide-open spaces. Because this environment seemed so like our own, Stone Age hunter-gatherers were thought to have had a lifestyle like those who existed in recent memory. For example, Marshall Sahlins, in the opening pages of his *Stone Age Economy* (1972), thought that '... on the evidence from modern hunters and gatherers... [the Stone Age] was the original affluent society'.

The idyllic savannah hypothesis could not explain the huge size of the human brain, which is costly in evolutionary terms and seems over-engineered for a supposedly relaxed savannah lifestyle. Archaeological measurements show that the brain enlarged as our ancestors formed larger groups and hunted cooperatively. But ganging up to bring down a mammoth does not require the brainpower and level of cooperation that can create a hydrogen bomb or build and fly a jumbo jet. Archaeologist Tim Taylor, for example, in *The Artificial Ape* calculates that if the cheetah had increased its speed in proportion to the human brain's improvement it would now be able to run at 200 mph, 120 mph faster than needed to catch its fastest prey.

The arithmetic behind the '200-mph brain' may not be very precise, but you get the idea: it is impossible for natural selection to evolve pointless luxuries. The brain is so powerful that it must have evolved to face challenges other than those facing the savannah-dweller. With hindsight, a clue existed in the Inuit craftsmanship on display in the British Museum. The innovative technology that these people developed in a few generations, to cope with a most un-savannah-like environment, shows the 200-mph brain at work. The Inuit colonisation of the Arctic happened less than 5,000 years ago, at least 45,000 years after the archaeological record shows that the human brain had stopped expanding. The Inuit had, and still have, the same brain as all of us.

To explain why a large brain evolved I will draw on the work of Professor Mark Maslin of University College London. His recent book *The Cradle of Humanity* (2017) is a guide not only to his own research but to all the latest discoveries in the field of human evolution. Maslin, like Kropotkin, is a geographer who has researched human evolution. My other major source is *The Evolution Of Our Species* (2012) by Professor Chris Stringer of London's Natural History Museum. Wherever possible I will rely on these two easily accessible authorities, quoting passages directly instead of paraphrasing. This will allow sources to be checked by searching the text, and will stop me distorting them with ignorant interpretation. I will indicate other sources by endnote references. My own insights will come from the fields in which I have worked, which are communications engineering and modern social history.

It was Maslin's research into Stone Age environmental instability that finally proved the savannah hypothesis to be as wrong as could be. Even bipedalism (upright walking) did not evolve on the savannah, but in areas of mixed forest, grassland, and water where it was advantageous to be able to hold onto tree trunks or to climb vertically to escape predators. The large and energy-hungry brain evolved after that, in a more demanding environment where nothing less than 200mph would do.

Climate change: we've been there

The ultimate cause of the human brain's rapid evolution was the weather, and the fact that our ancestor apes had strayed from the climatically stable equatorial forests where the 'great apes' lived. The very recent discovery of the role of a hostile climate in human evolution is important in the light of Kropotkin's research and theories. It was something undreamed of by Darwin, but clear to Kropotkin from his researches in Siberia.

Our knowledge of the climate endured by our early ancestors was dramatically increased by the Greenland Ice Sheet Project, which began in 1988. Drilling down through 110,000 years of ice formed by repeated snowfalls, the project reached bedrock in 1993 after bringing up a 3-kilometre-long ice core which contained a detailed record of the variations in the weather and the makeup of the atmosphere during that time. The information in this record astounded scientists. Analysis of the ice revealed that the climate before 12,000 years ago was hugely more erratic than previously thought, with icy and warm spells alternating within a few years, apparently due to perturbations in the earth's orbit.[7] The recent extended benign climate turns out to be unusual; so far it has seen 12,000 years of stability following a period at least a hundred times as long when the temperatures seesawed at a rate that makes the most alarming predictions of modern climate change look trivial. At the end of the short Ice Age known as Younger Dryas, when modern humans had already spread around the world, average temperatures rose by fifteen degrees centigrade in only ten years.[8] This would be like England turning into North Africa in a decade. By comparison, global warming increased average temperatures by less than one degree in the whole of the twentieth century. The most pessimistic prediction is for an increase of 6.4 degrees in the next 100 years, which would be catastrophic for present-day human society.

One professor of environmental science wrote lyrically of the new discovery of the prehistoric climate: 'some data are so sublime that they completely transform our picture of the world ... giving us a stunningly surprising picture of the sort of world in which our cultural system arose.'[9] The new Stone Age climate data meant that many 20th century theories of evolutionary psychology needed updating. As Dr Matteo Mameli of Kings College, London University, wrote in 2010: that 'Arguably, narrow-sense evolutionary psychologists often tend to underestimate the variability of hominin Pleistocene [1.8m to 10,000 years ago] environments and this leads them to simplistic views about the selection pressures that were operating on our ancestors at the time'.[10]

Apart from the significance for human evolution, this shocking weather record helped to kick-start the modern climate change debate, because it revealed that our planet's climate is fundamentally unstable.

Out of an African hell

The weather in which our species grew up turns out have been even grimmer than revealed in the worldwide data from the Greenland project, because our birth continent had its own special version of hell. Africa's modern torpid climate and geology bear no relation to the maelstrom that tormented it when it was humanity's cradle. Nowadays all the volcanoes, tsunamis, earthquakes and floods seem to happen on other continents, but for two million years they happened in Africa. Maslin describes how freak tectonic and climatic conditions due to precession (wobbling) of the Earth's axis of rotation made that continent resemble the set of an endless disaster movie. It was not a coincidence that *Homo sapiens* happened to be born in such a nasty place – the African climate created us.

Extinctions of early hominin (near-human) variants happened with such frequency that the fossil record shows that the holocaust was making natural selection work overtime. Having ventured too far from the more stable equatorial regions, the ancestors scattered in front of the storm of heat, wind, dust, and flood, trying to find safe havens. Using Maslin's chronology, the first early humans known as *Homo erectus* escaped to Europe and Asia about 1.7 million years ago. Others commonly known as Neanderthals and Denisovans followed them about 500,000 years ago. *Homo sapiens* first appeared in Africa 200,000 or 300,000 years ago and migrated into Europe about 120,000 years ago. By 40,000 years ago only *Homo sapiens* remained, both in and out of Africa, with some genetic material derived through interbreeding with other species including pre-human bipedal apes.

Stringer, in *The Origin of Our Species*, explains how for many years there was widespread belief that *Homo erectus* evolved into *sapiens* in many parts of the world after leaving Africa 1.7 million years ago. This theory has now been disproved, and *sapiens* is known to have originated only in Africa and then spread worldwide. Under the old theory, scientific attention had been focussed on that first 'out of Africa' migration. Now that it seems certain that our most advanced ancestors spent an extra million years trapped on that continent, the phrase '*recent* African origin' has taken over, and what happened 'in Africa' has become a new focus of research. That is why Maslin's research into the ancient African climate has become so important for human evolution.

Maslin's research reveals that earthquakes pushed up a bulge which cracked open to form the extraordinary structure which we misleadingly call the Rift Valley. It is not really a valley but a plain hung between two narrow three-mile high mountain ranges. The mountain walls acted as weather barriers creating a microclimate within the Rift. But these barriers switched on and off very suddenly as the violence of the weather outside rose or fell. The result was not one but many successive microclimates within the Rift as a combination of the outside weather and the 21,000-year precession cycle repeatedly flipped the mountain barriers on and off. This newly-discovered 'pulsed climate variability' in the Rift means that it would be better described as the 'womb' rather than its traditional description as the 'cradle' of humanity.

At some period the Rift became covered by a mixture of forest, grassland and lakes, and apes began to develop bipedal, upright-walking varieties better suited to this fragmented landscape. Then the microclimate switched to wet over a little more than a century; lakes inundated the Rift and drove the bipedal creatures out over the mountains to seek their fortunes elsewhere. Then the switch flipped back to dry and sucked some of the bipeds back in; a few thousand years later a gradual transition wiped out many of these. Extreme periods of variability occurred when the eccentricity of the earth's orbit was at a maximum. Each period corresponded to four or five lake expansions and contractions driven by the shorter precession cycles that cause the earth to wobble like a spinning top.

'The largest and most prolonged period of variable climate' according to Maslin, 'is between 2.1 and 1.7 million years ago', At that time, he says, 'hominin diversity reached its highest level' with two species generally considered of the genus *Homo* and two species of pre-*Homo* coexisting with each other. One of the former was *Homo erectus*, and he goes on to say that 'many of us would argue that the *Homo* genus should really start [1.8 million years ago] with *Homo erectus* given its similarity to modern humans and the significant differences with earlier *Homo* specimens'.

Among the new features of *Homo erectus* mentioned by Maslin are a brain 40 per cent larger than earlier members of the *Homo* genus and 80 per cent larger than earlier hominins. Other new features were 'adaptations required for long-distance running', 'shoulders that would have allowed the throwing of projectiles', a shorter gut, and males and females of similar size. In a fascinating section on sexual behaviour and human evolution, Maslin points out that the difference in size ('sexual dimorphism') is a good indicator of mating patterns. Gorilla males are twice the size of females and can consequently defend a harem-style mating arrangement. The sexual dimorphism of *Homo erectus*, however, was less than in earlier hominins and, according to Maslin, 'very similar to that of modern humans'. He cites this as an indicator of monogamy or serial monogamy, which is the pattern followed by most recent hunter-gatherer societies but is quite rare in other mammals.

Homo erectus therefore may have been the first human to adopt a (possibly serial) monogamous lifestyle. This 'pair bonding', according to Professor Robin Dunbar, does not (as commonly thought) indicate that the 'nuclear family' had been invented. In a paper[11] published by the British Academy in 2010 Dunbar showed that the principal payoff of pair-bonding (which he called 'besottedness') in humans is that it reduces sexual harassment and infanticide by males which would otherwise occur due to the large size of human groups and their habit of dispersing by day and only partially coalescing at night. Dunbar's calculations show that pair-bonding aimed purely at caring for infants could not have satisfied the high calorie needs of the expanded human brain in infancy. An early human infant must therefore have been receiving food from the rest of the group. Dunbar shows that another large increase in brain size seems to have followed pair bonding, probably because of the new cognitive requirements of falling in love. In some other animals, pair bonding is not symmetrical as it usually is in humans. This suggests to me that natural selection evolved the novel behaviour of mutually falling in love as a peaceful and group-friendly replacement for the disruptive mating violence of the great apes and earlier hominins. Involuntary mutual aid, in other words.

The meek inherit the earth

Reducing the threat of sexual harassment and infanticide by males reduces overall tension in a group. Threatening behaviour is common in animals groups which form a dominance hierarchy or 'pecking order'. It is commonly asserted that dominance hierarchies based on physical bullying reduce overt aggression by ensuring that everyone 'knows their place'. While true this ignores the fact that such hierarchies create tension within a group, and the *fear* of conflict which obstructs cooperative behaviour.

A demonstration that fear and aggression can block cooperation was shown in the experiment in which pairs of chimpanzees were given the opportunity to obtain food only if both pulled on a rope. They showed reluctance to do so when the food was easy for one chimp to monopolise and squabbling would result.[12] This shows how *fear* of conflict inhibits cooperation. Other experiments, inspired by the new discoveries about human cooperation, showed that chimpanzees can be far more cooperative than previously thought if conditions are arranged that neutralise their mutual hostility over food and sex. They quickly learned to identify which of their peers were better at cooperating. An animal doesn't have to be very smart to make the first cooperative move and then to learn from the beneficial experience.

The sexual dimorphism measure may indicate that the first cooperative human emerged 1.8 million years ago in *Homo erectus*.

Mark Maslin argues that *Homo erectus* may even have achieved the ability to share knowledge through speech — an extremely advanced form of cooperation. He bases this argument on early evidence of symbolic behaviour found in the site known as Trinil on the island of Java. There is possible supporting evidence from communications engineering, which I will discuss in the next chapter. The more widespread view is that language did not appear until *Homo sapiens*.

The evolution pump

The Rift Valley, where the new *Homo erectus* emerged nearly two million years ago, is part of a more extensive geographical feature running from Syria in the north-east to Mozambique in the south-west. According to Maslin, this linear structure caused the Rift Valley's 'pulsed climate variability' to become an engine for extruding the newly-formed and highly mobile, cooperative, and adaptable species into Europe and the rest of Africa:

> '… the occurrence of deep freshwater lakes would have forced
> expanding hominin populations both northwards and
> southwards, generating a pumping effect, pushing them out of
> East Africa towards the Ethiopian Highlands and the Sinai
> Peninsula, or into Southern Africa, with each successive
> precessional cycle.'[13]

These forced emigrants owed their survival to their new ability to survive in an environment different from that of their birth, and their long-distance walking ability. Bipedalism had evolved to suit the new vegetation long before the brain expanded in *Homo erectus*. The discovery of the skeleton known as Lucy (from 3.2 million years ago) provided clear evidence of this. Lucy was highly bipedal, but her brain volume was only 500 cc, the same as a modern chimpanzee and only half as big as *Homo erectus*.

Their legs allowed the variants formed and then expelled by the Rift's climate pulses to flee into the African hinterland, where they encountered every other variety of challenge to their mental flexibility. Africa is vast, comprising one fifth of the world's land

19

mass. For nearly two million years tiny populations of refugees fled to Africa's different regions.

The other apes were never foolish enough to wander into the Rift from their safe but geographically tiny and overcrowded homeland where the environment was not so variable. While their foolish *Homo* cousins let themselves be caught out in the open by the maelstrom, the ancestor apes continued to fight among themselves for the limited resources available in their stable but crowded habitat.

Sharing the Overkill

The 'group' that was so important in human evolution was not a band of hunters who came together to stampede a herd of wild cattle over a cliff. Such group activities were not typical of human hunting strategies during the two million years of evolution in Africa, though they became more common later on other continents. The default strategy contrived by natural selection was to hunt alone or in very small teams, because it is more efficient when the prey is hard to find. Once the spear was invented and the shoulders of *Homo erectus* became throwing machines, a single human had the ability to wound any mammal on earth and then track it until it collapsed. 'We are unique among primates in our capacity for endurance running,' according to Stringer, 'ungulates [hoofed animals] can run much faster than humans over short distances but completely exhaust themselves over long distances, at which point they are easy to dispatch.' Unobtrusiveness, endurance, and skill in tracking became the critical success factors, not numbers and brute strength. This is the case today in hunting societies.

The chancy nature of lone hunting means that the hunter sometimes returns to the group and his family empty-handed, depending for survival on another hunter having been successful and sharing his bag. This works better if each hunter is innately motivated to bring back more than was needed to feed his own dependants, i.e. overkill. Statistically, such an individual-based hunting strategy was certain to produce more dependable supplies for everyone in the group than going out mob-handed. One consequence is that the gut, the most energy-intensive organ after the brain, could become shorter. Another consequence is that each hunter's individual contribution to, and consumption of, group resources was evident to all.

Sharing can evolve under natural selection if unproductive individuals are prevented from reproducing as prolifically as those who take risks and expend energy in hunting. If 'free riders' could get more food with less effort than productive individuals, then any genes for over-productivity and sharing could die out.

Before a tendency to share could become innate and stable, therefore, there would also have to be an innate mechanism to identify 'free riders' and reduce their reproduction. In humans, this mechanism is self-advertising.

Advertising generosity

Human over-productivity in hunting began at least as far back as 400,000 years ago when the first evidence is found of big game hunting. Hunting big game made sharing necessary because it was too intermittently productive to feed a nuclear family.[14] Modern hunter-gatherers share a significant proportion of their prey outside the family, and research shows that this is related to their mating behaviour. Even when there are safer and more predictable sources of food available they hunt big game as a way of showing off their skill and generosity, and this apparently attracts mates. Kristen Hawkes, Professor of Anthropology at the University of Utah, argues that this 'unique male subsistence contribution may have evolved as hunting large animals became a focus of competitive display'.[15] It is a peaceable competitive mating behaviour evolved by 'sexual selection', like the peacock's tail, which evolved because female choice determines which male will reproduce.

Natural selection cannot evolve behaviour which reproduces other individuals more than it does the possessors of the behaviour, but human output-sharing is not a behaviour that evolved in isolation. It is part of a mating strategy that increases the reproduction of successful hunters more than their built-in generosity reduces it, so it can propagate through the species. It's as if the peacock's tail had been replaced by a sexually selected trait that was accidentally useful to unrelated individuals. (Women foragers today do not generally share their gathered produce outside the family, but this does not mean that they did not share their over-productivity in other ways. It's just that the archaeological record has shown us more about male production than female).[16]

Ostentatious generosity functions as a mating display in some other species, where it has been called the 'handicap principle' because it shows potential mates that the individual's genes have capacity in reserve.[17] In those species it tends to be a male-only behaviour which strengthens the dominant male's hierarchical monopoly of mates. Some potential recipients therefore refuse to accept the gift. It was not associated with male sexual hierarchy in early humans because of our nonviolent egalitarian pair-bonding. Human females are continuously sexually receptive, and their fertile period is concealed. The chances of a pair-bonded human couple having an infant were increased if they had more frequent sexual activity, so self-advertising would have been a 'spouse magnet' and not just a 'babe/hunk magnet'. It can increase individual reproductive success not by taking mates away from others but by increasing frequency of sexual behaviour in a pair-bonded couple.

This peaceful but competitive mating display aspect explains why humans do not surreptitiously leave their surplus food or their newly-invented tool on the outskirts of the camp for chance discovery and for the benefit of all, as modesty and Christian virtue might dictate.

It is fortunate that we should have adopted showing off as a peaceful way of getting sex, because our sharing became stable in evolutionary terms by using this feature to reduce reproduction in 'free riders' who don't have anything to share. Of course, evolution didn't plan to help unproductive individuals to survive (but not to reproduce) as a by-product of the human mating display, it just evolved by natural selection. Mutual aid, or generosity towards the less productive, is no more outlandish a product of sexual selection than the peacock's tail or the bull elk's forty-pound, four-foot wide bone hat.

Human hunting strategy evolved during two million years of confinement in Africa. There, the prey evolved in parallel to become wary of humans, making the stealthy lone individual the most efficient hunter. Later, after evolution had produced *Homo sapiens* which had quickly colonised new continents teeming with megafauna who had no fear of this insignificant-looking creature, group hunting became more efficient. Gangs of *Homo sapiens* slaughtered whole populations of animals by burning them out or by driving them into canyons or over cliffs. Nature did not have time to achieve ecological balance as it had done in Africa.[18] Humans drove native species to extinction before they had time to adjust. It is important to realise that this new style of hunting was not innate behaviour. *Homo sapiens* could figure out the optimum approach for new circumstances using his brainpower. We did not de-evolve our long legs, shortened gut, and shoulders adapted for throwing to adapt to the new situation. The default option of lone hunting remains hardwired in human physiology. Underneath our clever brains, we are living fossils.

The cooperation toolkit

During the 1.5 million years that separated *Homo erectus* from *Homo sapiens,* the human cooperation toolkit underwent many upgrades. These included a 50 per cent increase in brain volume compared to *erectus*, particularly in areas thought to be important to speech. This made the brain so large that, according to Maslin, 'mothers would have required help' to give birth (he gives a rather ungainly demonstration on YouTube). The airway also became more suitable for verbal communication. The outer portion of the eyeball (the sclera) changed to facilitate nonverbal communication. According to Chris Stringer:

> humans have an enlarged, unpigmented and therefore white sclera, which means we can detect where other people are looking; equally, they can detect where we are looking. This must have evolved as part of the development of our social signalling, enabling us to 'mind-read' each other

Then, during the first 150,000 years of *H. sapiens* on earth, a final change seems to have gradually taken place. Mark Maslin summarises the archaeological evidence that human pre-natal testosterone levels decreased. The result was that males became 'feminised', with less masculine facial structure and skeletal changes that in modern human populations and other primates are associated with lower aggressivity and reduced sexual promiscuity. These physiological changes were contemporaneous with the appearance of creative art in the archaeological record and this final version of *sapiens* is referred to as Anatomically Modern Humans (AMH). From 50,000 years ago *H. sapiens* looked and behaved like you, me, Galileo, or Florence Nightingale.

Maslin suggests that this last reduction in aggressivity was achieved through rough justice. He cites modern tribes in Papua New Guinea, where 'human proactive violence — that is, thought out, discussed, and planned violence — is used to curb, control, and cull reactively violent individuals.' So perhaps sexual violence was attenuated in the human phenotype (the behaviour pattern arising from the genes) by a massacre of alpha males – a revenge of the nerds using the newly-invented spear. In the striking phrase of philosopher Benoît Dubreuil, Stone Age humans were 'condemned to equality' by the group's ability to 'sanction' domineering members, a chilling reminder of the plot of *The Eiger Sanction*.[19]

It is significant that the reduction in aggressivity began when *H. erectus* developed a shoulder that could hurl projectiles. This would have made it unnecessary to engage an aggressive alpha male in close hand-to-hand combat.

There are other possibilities. Among the common chimpanzee's more peaceable close relative, the bonobo, according to Richard Wrangham, male bullying is controlled by females ganging up and thrashing the culprits. Wrangham advances a less violent possibility: that a cooperative human group acted to 'exclude the intolerant' members.[20] This might have led to natural selection against physical bullying if the ostracised bullies found life shorter and reproduction harder when expelled from the group.

If the more lethal sanctions were applied, the 19th century Russian anarchists' targeted assassinations of their torturers may have been a resurgence of a Late Stone Age behaviour. Assassination *within* the group would have been a more effective tool with which natural selection created cooperation than Darwin's proposed warfare *between* groups

What does this evolutionary trajectory imply for the human 'cooperative instinct', (or 'altruism' for those — *not* including anarchists like Peter Kropotkin — who believe in innate human benevolence)?

On the evidence of modern science, cooperation seems to reside not so much in a human 'instinct' as in a set of physiological adaptations that have left an individualistic human spirit trapped in a cooperative body. The individualist in us would not want third parties to see from our eyes that we are looking at an attractive member of the opposite sex (hence the mirror sunglasses sported by many a Lothario). A man might regret being feminised by natural selection, and envy those on the tail of the bell-shaped curve who still have the craggy, dolichocephalic features of earlier hominins. We may be offended that others do not appreciate us unless we 'show off'. It is irksome that our brains are so big that females need help to give birth. But these and other inconveniences are our inescapable heritage, even if they go against our individualist ambitions.

To quote Dubreuil again, we were 'condemned to equality' by nature. In later chapters I will show how a transitory hierarchical culture has managed to make today's humans feel unequal again.

Meanwhile, does the new research into the African paleoclimate show who has won the argument between Darwin and Kropotkin? Was our cooperative toolkit created by warfare or is it the result of natural selection exploiting mutual aid to cope with the freak environmental conditions? Did humanity's struggle for survival pit us against each other, or was it a two-million-year war against the gods?

The case for warfare

Darwin's conjecture that the 'never-ceasing wars of savages' were essential to the evolution of cooperation is accepted by some modern academics. Those who have published in favour of the theory include biologist E. O. Wilson (*The Social Conquest of Earth*, 2013); mathematician Martin Nowak (*Supercooperators*, 2011), economist Samuel Bowles (*Did Warfare Among Ancestral Hunter-Gatherers Affect the*

Evolution of Human Social Behaviors? Nature, 2009), archaeologist Tim Taylor in *The Artificial Ape* (2010) and even the science fiction writer Arthur C. Clarke in *2001: A Space Odyssey* (1964).

Modern supporters of Darwin's theory tend to also support the controversial theory of 'group selection'. The theory holds that natural selection might allow a group containing some individuals with fidelity and courage to reproduce better than a more selfish group even if fewer of the faithful and courageous survived to reproduce their self-sacrificing genes.

Samuel Bowles, whose 2009 paper in the prestigious journal *Nature* is relied on by some of the others I have named, quoted Darwin's passage and set out to demonstrate mathematically that 'altruism', as he called it, could evolve in this way as a result of intergroup conflict.

Bowles's method is a philosophical approach to explaining natural selection, supposing the evolution of a single mental process for cooperative behaviour. It does not consider how natural selection can produce the same behaviour without relying on unselfish genes, simply by combining biological traits in new combinations to produce perfectly selfish mutual aid. Natural selection achieved this in many other species, as Kropotkin showed, and even between species as with the small fish that eat the parasites that infest a larger predator's mouth. Bowles, an economist, may not be familiar with the human cooperative mutual aid toolkit that existed long before his evidence of warfare. The human non-violent mating, legible eye movements, shortened gut, tendency to show off, and feminisation of the male all condemned us to mutual aid despite our worst intentions.

On top of these, natural selection has our huge brain to work with on its artist's palette. Using that, it has made us dependent on involuntary symbolic behaviour that allows our fellow humans even to read our innermost thoughts. This symbolic behaviour is *speech communication.*

There is no suggestion that all humans possess the whole toolkit for cooperation. Natural selection can only go far enough to ensure that what it has produced will survive and reproduce, despite variations. Nor can natural selection ensure that what it evolved during the Stone Age will function in the same way in today's very different climactically stable environment. Nowhere is this more evident than in speech communication. That deserves a chapter on its own, which will explore its primitive origins and design and its shortcomings in modern hierarchical culture.

CHAPTER 3

The Human Internet

IN 1604 THE ASTRONOMER JOHANNES KEPLER gave the first explanation of the workings of the eye, using insights from his experiments with the recently-invented pinhole camera. The devices that we invent often help us to understand the natural world. When information technology arrived in the 1960s, psychologists tried to explain some characteristics of the brain by comparing it to a computer.

It's time to update that 1960s model of the human brain. A standalone computer is a poor fit with Robin Dunbar's discovery that neocortex size in early human increased as group size increased, to facilitate complex social behaviour. Dunbar and his colleagues have reflected that discovery in the title of their book: *Social Brain, Distributed Mind* (2010). The metaphor (or reality) of the human 'distributed mind' shows that we should compare a group of human brains to a distributed communications network. In a distributed network, multiple transmission paths link any pair of transmitters/receivers. A new transmitter/receiver can join the network at any point. This is the architecture of the internet. A group of speech-enabled early human brains was not a collection of computers. It was an *organic internet*.

The technical breakthrough that created the digital internet was the use of distributed computing power to establish communications between users connected to different computerised switches. Each switch can be connected to several other switches, which cooperate to find paths between source and destination. Before the internet, network architectures were 'hierarchical'; switches were typically connected to each other in either star or pyramid structures, and relied on a centralised authority to establish connections ('circuits').

The digital internet has shown that a network that does not need hierarchical authority can perform better. The word *anarchism*, meaning 'without ruling authority', could be applied to the architecture of the internet. It equally describes the organisation of a Stone Age human group, which functioned without hierarchy, as we saw in the previous chapter.

The invention of the Cloud

One aspect of the human internet that may not appear when language is considered simply as a means of thought or communications is its distributed information storage ability. With no other external storage medium (i.e. no writing) and no dictionary to record new words, the only way to update, correct storage errors and synchronise changing vocabularies was to continuously retransmit and refresh the group memory using speech.

Until long after the climate stabilised spoken language was the main way of accumulating and handing down culture and wisdom to future generations. As late as the 1700s, the illiterate Polynesians orally passed down accurate maps of the Pacific which enabled the explorer Captain Cook to draw charts for use by Europeans.[21] One can see how knowledge could become resident in the language by imagining a group that had forgotten how to make and use bows and arrows. One of them might say: 'I remember an old man who used to talk about sticks which flew through the air because of a piece of rawhide tied to another stick. What could he have been talking about?' A sequence of spoken symbols has then become not just a way of communicating knowledge, but a repository of that knowledge. Language became to the brain what 'the Cloud' is to a computer's hard disk: a backup master copy of the culture which is downloaded into other brains. Language and its group storage and correction capabilities justify the description by Dunbar and his colleagues of the human group's brains as a 'distributed mind'.

Researchers now agree that a stable culture emerged when local population reached a critical mass. Innovation has been studied in other primates (e.g. in adapting to new diets) and in computer simulation and it has been established that the net rate of innovation (creation minus forgetting) goes positive when a certain critical mass of population is reached, allowing innovations to accumulate.[22] Professor Mark Thomas of University College London summarised the critical mass theory at a Royal Society Conference on Early Anatomically Modern Humans in 2011: 'cultural sophistication reflects human interaction, not human intelligence.' In *The Origin of Our Species* Chris Stringer puts it another way: 'for the survival and propagation of knowledge, it's not so much what you know, but who you know, that matters.'

Creating the wisdom of the group

Scientists still debate why natural selection went to so much trouble to evolve human speech, when humans use speech to mislead as well as to inform. Some have even questioned whether language is the product of natural selection at all, proposing that it might have arrived as a single one-off mutation, for good or evil.[23] This hypothesis is hard to reconcile with the archaeological record which shows many incremental changes to human physiology that improve the ability to vocalise. Some of these changes are downright dangerous, such as modifications to the larynx which increase our risk of choking but improve human ability to vocalise. It has been suggested that natural selection could have evolved speech to deceive rather than to share useful information, a theory that cognitive psychologist Steven Pinker and others have criticised.[24]

The fact that speech is nevertheless often used to deceive highlights the advantage of living in a group and being part of a distributed communications network. The communicating group reduced the risk of deception by ensuring that a member did not hear only one version of events, but many.

Speech is commonly performed one-to-one (as in conversation) or one-to-many (as with animal alarm calls or media broadcasts). In an egalitarian social group, though, the most powerful form of communication is many-to-one. This is because each receiving brain is connected to the source of new information by multiple independent communication paths passing through different individuals. If humans routinely repeat what they have heard from other members of the group (as they do, in gossip) they create a multipath communication network: an internet.

A multipath communications network has inherent error-correction properties that can explain how human speech began and how it stored accurate knowledge in the Cloud. More importantly, these properties highlight the limitations of human error-correction and thought processes. The consequence of those limitations are for later chapters. This chapter explains how human language originated because of the group's internet architecture and why that evolution gave rise to 'vulnerabilities' as they are called in modern computer jargon.

The error-correction mechanism most easy to apply in a multipath network is a 'majority vote' in which the recipient accepts as true the data that arrives most often, and rejects less frequent variants as errors. In communications engineering this multiple transmission and balloting as a means of correcting errors is called 'repetition coding'. It is an example of the more general technique of 'redundancy' – transmitting more symbols than necessary so that they can be checked against each other for consistency. A Stone Age group member heard many 'redundant' versions arriving through different transmission pathways in the group. It does not take much brainpower to use the most common version.

As an example of the power of repetition coding, imagine that an early human was able to ask: 'are there snakes on that hillside?' If everyone in the group had an error probability of 20 per cent and if the questioner asked only one individual the chances of getting the right answer would be only 80 per cent. If she asked five individuals and took the majority vote the chances of getting the right answer would increase to 94 per cent. Robin Dunbar's well-known estimate of the size of the *Homo sapiens* group is 150. If the question was asked of only 30 of those and they each had the same 80 per cent chance of telling the truth, the poll would provide the right answer in 999 cases out of 1000.[25]

The above explanation is based on communications engineering, not biology. This cross-disciplinary approach may be a point in its favour. Stephen Pinker says that:

> good theories [of biological adaptation] use some independently established finding of engineering or mathematics to show that some mechanism can efficiently attain some goal in some environment.[26]

If 50,000 years ago the human brain used the repetition coding mechanism to correct speech transmission errors, we may still be using it today. Experimental evidence of this would go a long way to establishing the theory. There is such supportive evidence in numerous experiments carried out since the 1970s.

The landmark experiment was that of Hasher, Goldstein, and Toppino at Temple and Villanova Universities in Pennsylvania. In their paper *Frequency and the Conference of Referential Validity* (1977) they showed that simple repetition of plausible but false statements was enough to make people rate them as true. This phenomenon is now called 'the truth effect' or sometimes 'the repetition-induced illusory truth effect'.

The experiment used statements of plausible trivia unlikely to be within the subjects' knowledge. The statement 'The capybara is the largest of the marsupials' is unlikely to trigger any cultural prejudices. If repetition makes it appear true (it is false), it is likely to have something to do with nature rather than nurture, with evolution of the brain rather than with acquired culture.

Christian Unkelbach of Cologne University explored the possibility that the truth effect serves a useful purpose. He presented his paper *Gullible but Functional? The Role of Information Repetition in the Formation of Beliefs and Values* at the 2017 Sydney Symposium on *Social*

Psychology of Gullibility. Unkelbach summarised all the suggested explanations of the repetition truth effect and asks whether it is 'functional' by which he means that it leads to more true beliefs than false beliefs. He concludes that it does lead to more true beliefs, with the important reservation that 'strategically false' communications can exploit repetition-induced truth to spread false ideas:

> Thus, inferring truth from repetition may be an easy and useful shortcut to adequate truth judgment. However, in cases of strategically sent and repeated false communications, inferring truth from repetition comes at the cost of sometimes false beliefs.

Do these experiments support the claim that repetition coding was the origin of successful speech transmission in the Stone Age? This is where evolutionary psychology can provide an answer. The idea of evolutionary psychology is that the brain finished evolving at least 50,000 years ago and must contain specialised functions that solve problems that existed before then but not problems that only arose afterwards. An example often quoted is the craving for sugar. It was beneficial when sugar was a rare energy supply for our infants' hungry brains, but now that it is common the craving for it is a health hazard.

The important *evolutionary* question is not, therefore whether the repetition-induced truth effect is doing more harm than good today. Evolutionary psychology instead asks the question: *did the repetition effect transmit the truth, and only the truth, during the period of evolution of the human brain?*

Surprisingly enough, the internet analogy shows that the answer is a resounding *yes.* That is because in the leaderless anarchist group there were no privileged information sources broadcasting 'strategically sent and repeated false communications'.

It should be of deep concern that human society and culture depends on such a primitive and easily hacked communication system. Fortunately, repetition coding is not the only error correction system that we have inherited from early humans. Humans have a way of detecting bias, but only if they are trained to use it.

To be biased is human

Error correction will be possible if the recipient can detect when information received has been distorted in transit. Distortion is most likely to arise from the thought processes of the transmitter — bias, if you like. One of the most powerful features of the human brain is its ability to detect bias, provided it is intensively trained in the same way as it is trained in childhood to use speech. It works by deducing the speaker's mind-state.

Robin Dunbar has explained that his 'social brain' hypothesis predicts that humans may be able to read each other's minds. The social brain hypothesis:

> 'principally focuses on the ability to use knowledge about other individuals' behaviour and perhaps mind-states to predict and manipulate those individuals' behaviour … [when] managing the social relationships on which … day-to-day survival and reproduction depend.'[27]

Knowledge of other individuals' 'mind states', known as 'Theory of Mind', or 'mentalising', is what enables humans to detect bias. Dunbar rates its evolution in primates on a scale from 1 to 4. Level 1 is awareness of the state of one's own mind. Level 2 is the ability to perceive the mind-state of another individual, and is necessary for the simplistic correction of errors by repetition coding as described earlier. Chimpanzees can operate at Level 2. For example, a lady chimp wanting to groom with a young male will do it behind a rock with only her head visible to the 'big daddy' chimp who thinks he owns her. This shows that she has quite an acute understanding of what's going through the dirty old man's mind and of what he can see. *Homo sapiens* can operate at Theory of Mind Level 4 or even higher, enabling a human to guess what another individual has deduced about the mindsets of people at 2 and 3 removes. Early humans appear to have steadily evolved from level 2 to level 4 over time, judging from archaeologists' measurements of the volume of the frontal lobe where there is evidence that the Theory of Mind resides.[28]

An individual with basic chimpanzee-level (Level 2) intentionality knows that not everyone has the same mind and will, if able to execute the 'majority vote' algorithm, decide what to believe based on a simple majority. Each subsequent 'level of intentionality' improvement is a game-changer. The statement 'Sally's mother says that Sally is a genius' allows a level 3 listener to infer that the speaker intends the introductory phrase 'Sally's mother says …' to be an important qualification of what follows. This is because the level 3 listener can imagine what the speaker thinks that Sally's mother's prejudices are likely to be, and so will not necessarily give extra weight to any information received from Sally. For an individual who is level 2 the introductory four-word qualification does not play any part in decoding the message. This means that any individual who relays the entire sentence is accidentally providing more helpful information to level 3 listeners than to level 2s. Level 2s will be disadvantaged because they have heard that Sally is a genius and they will give extra weight to her opinion that there will be no snakes on that hillside.

This discrimination against individuals with lower level of Theory of Mind shows how the brain's language capability evolved to meet the selection pressure imposed by other brains. If your brain was an old model and you couldn't understand the most advanced speakers, your prospects of passing on that old model brain were diminished, just as a bantamweight male gorilla stands little chance of passing on his slender build.

Culture goes into meltdown

Theory of Mind can, in theory, protect humans against false beliefs created by the brain's basic repetition coding mechanism. In practice, natural selection designed it to work in relatively small groups who knew each other well enough, directly or indirectly, to practice a bit of psychoanalysis. In a more complex written culture, a scholar can sometimes deduce the mindset of an author whom they don't know personally, allowing them to dismiss some judgements of an otherwise respected writer. That's how Kropotkin was able to see that Darwin had written passages under the influence of Malthus's highly political obsession with population control. He could also deduce from his own wide reading and experience that Darwin, like Malthus, had no direct knowledge of harsh climates or sparse animal populations.

Only if humans have a wide experience of life and a forensic, analytic approach to information can they avoid falling victim to today's repetitive mass media including electronic robot propaganda ('strategically sent and repeated false communications' in the polite words of Christian Unkelbach).

Yuval Noah Harari's view articulated in the first chapter therefore seems to be correct in one regard. Knowledge sharing through speech, the most powerful tool of human cooperation, 'evolved for millions of years in small bands of a few dozen individuals.' It did not evolve to deal with mass media, particularly the 'strategically sent and repeated false communication'. But Harari's conclusion does not seem to be correct. He believes that 'hierarchies' of the type that exist today have ensured cooperation by overcoming the shortcomings of evolution. But they clearly have not overcome the deadly shortcomings of the repetition coding method of error detection., as the experiment of Hasler and colleagues showed.

The stated purpose of this book requires an answer to the question: could ruling hierarchies including the state be *causing* the modern limitations of mass cooperation? That's why the next chapter will examine the impact on knowledge creation of the cities, kingdoms and empires that followed the Agricultural Revolution.

CHAPTER 4

The Eden Doctrine

H UMANITY'S FINAL ESCAPE FROM AFRICA took place around 55,000 years ago. The human population had begun to increase dramatically through the critical mass effect which ensured that life-prolonging knowledge accumulated from generation to generation. If humans had become so smart by then why did it take them until 12,000 years ago to begin their career in agriculture? The answer to this question was provided by the discovery of Stone Age climate change: organised agriculture was simply not possible until then.

The climate stabilised around 40,000 years after the global migration began. Temperature, greenhouse gas levels and humidity increased sufficiently to allow intensive agriculture. Humanity was on the job almost immediately in several different locations.[29] Long-separated groups of humans almost immediately worked out how to modify local wild plants by selection for optimal food yield. Wheat predominated in East Asia, rice in China, maize in Central America, the potato in South America, and sunflowers in North America. Humans were just as quick to breed local animals to improve their meat, milk, or pulling power. It was a huge change for the human species which had evolved as hunter-gatherers, omnivores surviving on wild plants as well as animals we could scavenge or kill.

This first Agricultural Revolution is a testament to the winning talent of our species: the extraordinary mental adaptability which had previously enabled us to reproduce in the most unstable environments. Knowing what we do about this adaptability, there is an obvious problem in explaining why we did not rush headlong into the Industrial Revolution immediately once agriculture provided capital. The answer lies in the peculiar thought-controlling hierarchical ideologies that sprang up around agriculture. Our best example is the story of the Garden of Eden, designed to stop people becoming hunter/gatherers.

Forager versus farmer

Many human groups refused to adopt the agricultural way of life. After the climate stabilised around 12,000 years ago and farming began, life for those who continued as hunter-gatherers became a walk in the park. No longer was it necessary to either migrate or learn how to survive in a new environment every few generations – savannah in one generation, tropical rainforest a few generations later. Anthropologist Hugh Brody, who has compared advanced hunter-gatherer societies to agrarian cultures, points out that, even before farming, 'hunter-gatherer populations in other more temperate places must have increased with relative ease.'[30] How pleasant life must have been for a hunter-gatherer in the Thames valley in the stable temperate climate of the late Stone Age without a gamekeeper or irate farmer in sight.

It is true that as the farmers expanded, the hunter-gatherers were gradually pushed back into regions unsuitable for farming like the Kalahari Desert or the Arctic but even in such places life became more than bearable. The newly-stable climate allowed the hunter-gatherers to accumulate and refine local knowledge generation after generation. The last Kalahari San foragers only had to spend a couple of hours every day searching for food in a region where we would die of starvation or dehydration within days. The average workload for all known recent hunter-gatherer groups is four hours per day.[31] In one of Brody's documentary films a San reminisces about a 'big shop' which seems to have involved the appropriation of the fat

38

store accumulated in the tissues of the bat-eared fox, then a walk of tens of miles to obtain a particular wild melon which could be stuffed with the fat to make a succulent dish which would last for weeks. These recent hunter-gatherers were accustomed to gourmet meals in the desert, not iron rations.

The recent foraging or hunter-gatherer societies were 'advanced' in having superior notions of herbal medicine, which was practically all that the world had until 1850. They studied with great attention the habits of wild animals and plants. They did not separate animals and plants into useful and otherwise – the Inuit language, for example, has no word for 'vermin' or 'weed'.[32] They had no writing – that came with farming – but they had religions which explained to them why they lived as they did and where they fitted into the cosmos.

The agricultural way of life was not as easy as foraging. The archaeological record shows that settled farmers were shorter and less healthy and died younger than their hunter-gatherer predecessors. The reason lay in their poor living conditions; they were surrounded by herds of domestic animals which were a source of epidemic disease, camped permanently on top of their own effluent and that of their livestock, and living on an unhealthy diet often consisting of porridge three times a day. Paradoxically, though, their numbers increased dramatically because the sedentary life allowed a woman to raise several infants simultaneously. The more nomadic hunter-gatherers had used various forms of population control; as one authority puts it: 'The Australian aborigines killed both boys and girls, in order to preserve the mobility of the mother'.[33] In other more resourceful societies they may have employed longer weaning to reduce fertility and abortifacient drugs (which were well understood in antiquity).

The high birth rate of farmers explains why they expanded so fast, but the question remains: what was their motive for remaining in agriculture when life was so much easier and healthier as foragers? And if population growth caused the best farming lands to become overcrowded, why did the surplus population not revert to foraging? A partial answer to the second question is that they often did. As recently as the nineteenth century some Native American tribes left agriculture and returned to the plains to hunt. The fad for mass slaughter of 'game' animals by the British aristocracy in the name of sport in the same era was encouraged by its effect in preventing lazy labourers from reverting to an easier hunter-gatherer lifestyle as poachers.

Any explanation of the original development of agriculture also needs to account for the fact that intensive farming techniques began not in the most fertile valleys but in the less fertile uplands.[34] To judge from the first written narratives, hierarchical pressures encouraged a continuous expansion of farming into less promising lands. The Old Testament is one such narrative, which promotes an authoritarian materialism of a type not seen in hunter-gatherer societies but familiar in our own.

The first instruction manual

There must be few people in traditionally Abrahamic (Islamic, Christian or Jewish) societies, which together make up more than half of the world's population, who were not taught when very young the story of Adam and Eve being expelled from the Garden of Eden. The first version of this story was created in Sumeria, the first society known to have practiced both intensive agriculture and writing, in about 1800 BC. This Sumerian version describes the gods forming a woman from the rib of a man, as in Genesis. The story was rewritten in Hebrew in around 1000 BC, by someone known to scholars as "J" who was not a religious writer and was probably a woman.[35] Around 450 BC a Jewish priest, thought to be the prophet Ezra, combined it with other material to create the first five books of the Old Testament. The Qur'an contains a similar story. In the Jewish and Christian version, God puts a curse on Adam and Eve for their disobedient consumption of the fruit from the Tree of Knowledge, and they are expelled from Eden. Adam is condemned to cultivate hitherto barren land, earning his keep by the sweat of his brow, while Eve is punished by having to bear children in pain. Neither disobedience nor punishment is recorded in the earlier Sumerian fairy tale. The original of J's non-religious version is now lost, but it seems unlikely that this punishment was a feature of her bedtime story; most likely it was added by a priest such as Ezra.

It may come as a surprise to learn that forager societies have no Eden myth or anything like it. Only agricultural societies are encouraged by myths and religions that portray hard work and suffering as humanity's destiny, and this may be a clue that civilised society was keen to keep its early members from reverting to an easier life. Hunter-gatherers create stable and sophisticated societies without yearning to expand their territory or their population. This fact may seem surprising to those who have been presented with a model of hunter-gatherer based on the nineteenth-century Plains Indians who are reputed to have judged a man's worth by the number of scalps from other tribes sewn onto his shirt. This seems to betoken an expansionist philosophy like our own, if a more primitive one, and has reinforced the mistaken belief that the urge to conquest is innate. Our expansionist myths, such as the story of the expulsion from Eden, have been misinterpreted as signs of this supposed underlying genetic imperative. What our genes would say if they could talk, what they did say when we first began to talk. *'Go forth and multiply ... the sweat of thy brow ... dominion over the earth,'* ... that sort of thing. Not so, it now appears, because the hunter-gather groups are no different from us in genetic makeup and they have no such imperative myths.

An agnostic's belief that although Genesis is not literally true there must be a truth underlying the fable of Eden is not even a partial enlightenment: it only reinforces the myth. Philosopher John Gray thinks that 'the myths of religion are ciphers containing the truth of the human condition.'[36] James Lovelock, an eminent scientist and ecologist, suggests that the Eden myth corresponds to a genetic truth, writing of 'our inherited urge to be fruitful and multiply and to ensure that our own tribe rules the Earth'.[37] On the contrary, an examination of our long pre-agricultural existence as hunter-gatherers shows us that there is no more truth in the Eden myth than there is in the Highway Code. Far from corresponding to our human condition, Genesis is a set of instructions for adopting an alien way of life and creating an inhuman economy.

The historical significance of the Genesis myth is as a marker for when and how the change occurred. As Brody puts it: 'the truth of Genesis lies in the profound and disturbing insights it offers into the heart of society and economy that came with – and descend from – agriculture.'[38] Despite such disturbing insights, this book is not a plea for a return to a romantic hunter-gatherer existence. Any sentimental admiration should be reserved not for a vanished lifestyle but for that remarkable assemblage which we have inherited, the Stone Age human brain.

Onward and upward

Agrarian religions demand cultivation of less fertile lands; the easy lands needed no such propaganda. The Mojave Indians living along the Colorado River simply had to push melon seeds into the mud with their fingers to raise crops at any time of year, and had no need of complex husbandry, storage, or irrigation for their agriculture. Similar practices were followed in what we now call the Middle East, at one time an extremely fertile area. Only when the population in the most fertile valleys reached saturation point would their descendants have to be persuaded to use the less-productive uplands instead of abandoning agriculture.

The account in Genesis of the creation of mankind and the curse that followed is not descriptive. It is *prescriptive* because it tells man and woman what God (or natural selection, if one believes the Gray/Lovelock secular version of the myth) created them to do. The message of the story is that man has been designed to work hard to cultivate virgin soil, rejecting the easy life of the hunter-gatherer, and that woman is designed to bear children despite the pain (unique to humans because of the large cranium to accommodate the expanded brain), rejecting the hunter-gatherer's easy option of family planning. The *hadith* (sayings of Muhammad) contain a similar command to 'procreate and abound in number.'

In the Old Testament the Jewish God instructs mankind to:

> Be fruitful, and multiply, and replenish the earth. And the fear of you and the dread of you shall be upon every beast of the

earth, and upon every fowl of the air, upon all that moveth upon the earth, and upon all the fishes of the sea; into your hand are they delivered. Every moving thing that liveth shall be meat for you; even as the green herb have I given you all things.

Thus humanity is told that the whole planet and all its resources are created for their use. It was the Creator who decreed that things should be that way, and who had decided that man shall live by the sweat of his brow while the woman suffers in childbirth. It was a statement that the listener's lifestyle was pre-ordained, innate. It is remarkable how often scientists secularise this myth closely by replacing the Creator by a nearly identical supposed 'genetic inheritance'.

Agriculture wasn't pre-ordained, of course. It began by chance. It's inevitable that foragers who dropped seeds on the ground would observe the result, given the new stable climate. But as with many historical questions, too much emphasis has been placed on asking why agriculture began rather than why it continued and intensified. Similarly, some historians seek the reasons for the carnage of World War I by asking how it began, and arguing about whether Germany was to blame for the outbreak, as if World War I was a complete package waiting either to happen or not happen. It would be more illuminating to ask why the war *continued* so long and pointlessly and reached such heights of destruction. Many more factors are needed to account for that and many states are responsible. There are many reasons why agriculture continued and intensified despite its health disadvantages, tedium, frequent famines, and vulnerability to bandits. Among the reasons are that it was easy to teach by rote, the children could be put to work as soon as they could walk, and that it lent itself to, and supported, hierarchical organisation.

Organisation man arrives
It was the feasibility of storing the surplus that made it possible to introduce an authoritarian hierarchy. The surplus enabled some individuals to exist without working the land, and specialists could

develop full-time occupations other than farming. One of these occupations, to judge from the large numbers of surviving astronomical calendar stones, was the calculation of time and the seasons. Another, most probably combined with astronomy, was the priesthood. These specialists lived on part of the surplus produce which they extracted as tax. The surplus could also support a standing army that could easily enforce payment by settled farmers. The mechanisms of the coercive state had begun to fall into place and to become stable.

Agriculture introduced humanity to a new form of property, different from the *intellectual* property that hunter-gatherers created so prolifically. An agricultural surplus is a *tangible asset;* intellectual property is an *intangible asset.* Intangible assets cannot feed the upper layers of a hierarchy, which is why the new property-intensive hierarchical societies had no great interest in intellectual property. Even today, governments ignore most intellectual property when calculating industrial capacity using the GDP measure.[39]

In early agricultural societies the stability and growth of the asset-hungry hierarchy depended on the farmers' readiness to keep their noses to the grindstone. It is hardly surprising that writers and storytellers among the specialist elites put a motivational spin on their ancient tales. That's how the priest Ezra embellished J's story of Jehovah in about 450 BC. In J's original tale Jehovah was a mischievous and approachable spiritual companion; to have an idea of how foragers perceive such gods it is better to compare them to our imps, leprechauns, and wizards rather than to the 'sky gods' who developed to oversee intensive agricultural societies.

The sky god promised eternal life after death to make up for the farmer's life of struggle and pain on Earth. He told us, through the priests, to be fruitful and multiply, to breed cheap labour for the farm and to reject abortion and infanticide, to expand into barren lands and make the desert bloom, to regard all plants and animals as our servants, to forget about all other gods, and incidentally to give some of the surplus to the priests to feed themselves and the needy. Ezra incorporated the new motivational version of J's old Eden story into the Book of Genesis which is the first part of the Torah, a Hebrew word meaning 'instructions' – literally a user manual for the planet.

The 'wheat in the barn' accumulated by following those instructions enabled society to fund other projects. The Book of Genesis goes on to approve the success of Cain, the crop-farmer and later city builder who could do no wrong in God's eyes after he had killed his less economically productive brother. City-building and industry depend on the riches provided by agriculture and the mindset that comes from learning to control the environment instead of adapting to it. The command 'be fruitful and multiply' required prolific breeding, hard work, conquest, and conversion of new territory to agriculture.

Genesis: cause or effect?

It is tempting to believe that whole populations fell captive to a confidence trick practiced by their weather forecasters and priests when the latter's charitable tax-gathering schemes spiralled out of control and condemned us to enslavement by a wealthy state. This would be compatible with James Lovelock's statement that humans are 'belief engines' whose brains are incapable of scientific reasoning.[40] Lovelock's observation seems at first sight to be an odd claim from a scientist because it seems to say that what we have called science could never have happened. But, as we saw in the last chapter, humans do believe what they have heard repeatedly without any evidence, and Lovelock's 'belief engine' view of humanity is to that extent true. Is it the whole truth? Are we entirely under the control of the contagious ideas that we have 'caught' like a disease, from a religious culture that is like a computer virus evolving under Darwinian natural selection using our brains as its host?

But it is hard to see scripture as simply a rogue meme that has enslaved us. For a start, any reasonable sceptic must harbour the suspicion that it has never been believed by more than a tiny fraction of those who repeat its teachings. The benefits of pretending to believe were considerable and the chances of being exposed as a secret agnostic were negligible.

The second reason why the Old Testament farming ideology cannot alone be responsible for the growth of state power is that it wasn't only in the region we know as the Middle East that the expansion of agriculture took place. We see it in distant and isolated agricultural societies, for example the Incas of South America and the Bantu of Africa. Those groups did not develop advanced technology and sophisticated writing at the same pace as the Eurasian agriculturists, for reasons of geography and climate which Jared Diamond has explained. But Bantu and Inca succeeded in expanding beyond their homelands and converting much of their respective continents to agriculture, displacing the hunter-gatherers. This shows that that there must have been some non-Abrahamic influence at work favouring agricultural expansion. Each agricultural society had its own cosmology that taught them that their urge to expand was innate, erasing their hunter-gatherer heritage as completely as did the Torah in Europe and Asia. Not as much is known about these myths as about the Eurasian ones, but for example the Bantu code made unrestrained breeding obligatory.

The fact that similar mythologies have sprung up independently in at least three separate continents indicates that the Old Testament may be an effect, rather than a cause, of the first Agricultural Revolution. The primary cause was a climate which made taxation possible and hierarchy, possibly the favourite organisational structure of the universe, inevitable.

We saw earlier that the innate dominance behaviour that affects other great apes had disappeared in humanity when *Homo erectus* adopted nonviolent mating 1.8 million years ago. There is no evidence of hierarchical authority relationships during the evolution of humanity once the brutal mating dominance stage had passed. Recall the striking phrase of philosopher Benoît Dubreuil who said that Stone Age humans were 'condemned to equality' by the group's ability to 'sanction' domineering members. As soon as stored agricultural produce could support a standing army to defend dominant humans, hierarchy became stable, this time a product of culture rather than of a genetic predisposition to fight for more sex.

46

Cultural hierarchy has even allowed us to recreate the male gorilla's harem. Mark Maslin points to studies showing that although 84 per cent of 184 different cultures engage in polygamy, most males in those cultures are monogamous. 'It is usually only high-status or wealthy men who are likely to have polygamous marriages ... they can protect multiple females, usually by paying other men to protect them.'[41]

The singularity of human freedom

Humans are the only species to have exploited the storage of food without adapting genetically to a hierarchical organisation. Ants and bees developed genetically subordinate castes. Humans have not done so but instead have adapted even more successfully to hierarchy's needs, by cultural indoctrination.

The singularity of human agricultural adaptation reveals something uncanny about the timing of events during human evolution. The coincidences might give comfort to a creationist who believes that an intelligent designer planned the Earth's geography and weather and hence predetermined the cooperative human physiology that inevitably followed.

First, there is the curiously cyclic Rift Valley climate. In conjunction with the vastness of the inhospitable surrounding African land mass, this formed a giant engine for producing a species capable of creating knowledge and transmitting it down the generations. Then there is the fact that the weather became stable just *after* humanity had become capable of knowledge formation. If the climate had stabilised a million years earlier, the human brain would never have reached the stage where it could adapt to changing environments. Humanity would not have become an anarchist cooperative. We might have developed into giant ants, with different genetically-determined castes suited to the new agricultural existence.

On top of all the coincidences there is a distinctly New Testament flavour to mutual aid, and the 'distributed mind' seems to validate the Buddhist concept of the non-existence of 'self'.

The recent discovery that the monster brain and the store of symbolic knowledge evolved due to an extremely improbable combination of climate and geography has implications for the search for extra-terrestrial life. We may find millions of planets supporting life before we find one whose inhabitants have a brain anything like ours.

Impressive though it may seem, the human brain and its biological 'Cloud' of stored knowledge was, as we saw in the last chapter, haphazardly cobbled together and error-prone like other biological systems. It suffers from the major weakness that when group communications becomes hierarchical, the correction of errors through repetition coding and mentalising breaks down. The next chapter uses written historical sources to describe how hierarchical control plunged humanity into the Dark Ages.

Some historians identify the collapse of the authoritarian Roman Empire as the cause of this 900-year pause in knowledge creation. Human psychology guides us to the more scientific view that it was the thought-controlling legacy of the Roman Empire that turned off the lights, as we shall see.

CHAPTER 5

The Human Internet Fails

J ULIUS CAESAR HAS LEFT US A USEFUL SNAPSHOT, in his *Gallic Wars*, of our ancestors at an intermediate stage in the transition from egalitarian foragers to hierarchical agriculturists. He also lets us see how imperialists like the Romans coerced the tribes into abandoning their egalitarian ways.

Agriculture was not yet widespread in Northern Europe in Caesar's time. Three centuries later, after the Romans withdrew from Britain in 410 AD, the inhabitants slumped back into a hapless existence as victims of every foreign tribe that wanted to try its luck at conquest. Eventually the British enlisted pugnacious tribes from north Germany as settlers to defend them against unwelcome visitors, and the result was that Angles and Saxons colonised the country by invitation half a millennium before the Norman Conquest. Caesar had met and fought these tribes several centuries before their migration to Britain, at a time when the Roman Empire was still expanding and encroaching on German territory. He describes in his memoirs how the tribes had not yet adopted agriculture and were opposed to it for reasons which have proved far-sighted:

They have neither priests to perform sacred offices, nor do they pay great regard to sacrifices. They recognise only the gods whom they can see, and who are useful to them, namely the sun, fire, and the moon; they have not heard of the other gods even by report. Their whole life is spent in hunting and in pursuit of the military art; from childhood they devote themselves to fatigue and hardship ... They do not pay much attention to agriculture, and a large portion of their food consists in milk, cheese, and flesh; nor has any one a fixed quantity of land; the leading men each year apportion to the tribes and families, who have united together, as much land as necessary in the place in which they think proper, and the year after compel them to remove elsewhere. For this custom, they advance many reasons – they fear that they may exchange their ardour in the waging of war for agriculture, or that they may be anxious to acquire extensive estates and the more powerful drive the weaker from their possessions, or that the desire of wealth spring up, causing divisions and discords; they see that they can keep the common people in a contented state of mind when each sees his own means placed on an equality with the most powerful. ... It is the greatest glory to the several states to have as wide unpopulated areas as possible around them, their frontiers having been laid waste. They consider this the real evidence of their prowess, that their neighbours shall be driven out of their lands and abandon them, and that no one dare settle them; at the same time they think that they shall be on that account more secure, because they have removed the fear of a sudden invasion.

Welcome to a crowded world, where tribes had learned to defend themselves not only from their neighbours but also from rampaging imperialists like Julius Caesar who had a vested interest in converting them to agriculture because of the extra tribute it made possible. The tribes were right on target in their fear that agriculture would destroy their egalitarian society. They didn't foresee the worst of it – that the new hierarchy would, in the name of stability and security, make the sharing of knowledge illegal.

The cultivation of inequality

The first attempts at agriculture in Britain used an open field structure to continue the egalitarian traditions. Families were awarded several strips of land of around half an acre, located at different places in a jig-saw puzzle of such strips; this ensured that the whole lot could be ploughed by draft animals owned in common, and that each family would get an equal share of both the richer land and the poorer. Cattle were kept on common ground. No attempt was made to improve the fertility of the land.

As the system of agriculture spread, settlements became attractive to raiders, and a 'manorial' system evolved in which a local chief took responsibility for defence and taxed the ordinary people to pay for it either in produce or in service. This arrangement was less violent and disruptive than being raided by opportunistic passing bandits. Today the arrangement is sometimes referred to in a complimentary way as the 'stationary bandit' system, or kleptocracy, and is often deemed to be the foundation of all civilisations.

As Britain became more peaceable and unified after the Norman Conquest, the emphasis of the Lordship of the Manor (and the monastery, which played a similar role) turned from purely defensive to economic: the possibility of increasing production to generate more tax attracted attention. This required experimentation, and the open field layout where everyone sowed and harvested the same crop at the same time and mixed their herds on the common land did not lend itself to innovation in crop or livestock management. From the thirteenth century onwards the trend began to abolish open ground by enclosing it in parcels: privatisation, we would call it today, and the productivity of the land rapidly increased as innovative owners tested different types of crop rotation and cattle breeding. The enclosure of land was a positive development as far as productivity was concerned. Most collective farming schemes have failed because they have been centrally planned and unable to deal with the innate human need to innovate and to receive public recognition of their individual efforts.

Although it improved productivity, enclosure increased inequality. Displaced users of the commons had to turn to labouring for wages to survive. In view of the increase in productivity it should have been possible to share the gains from the new system to everyone's benefit, but sharing was no longer as simple as when the hunter-gatherer brought home his prize to be shared among the group in full sight. There were now layer upon layer of rulers to be maintained in increasing degrees of luxury, the topmost of which had a liking for expensive foreign wars. For a long time, exports were favoured because the rulers believed that an inflow of money was the only criterion of economic success, and this globalisation distanced even further the producer from his beneficiaries.

Merchants and intermediaries participated at every step, and the more calculating part of everyone's brain – the cerebral swindler who could see ways to overstate their value to the group – found opportunities in what we might now call the 'smoke and mirrors' of the new agricultural economy. It was impossible to distinguish the parasites from those who were doing an honest job of distribution. Dependence on a few crops also made the whole system susceptible to famine and even in good years significant numbers were starving to death or dying of diseases aggravated by malnutrition. In years of bad harvest grain prices rocketed and there was evidence that farmers, millers, and dealers took advantage of the price inflation to hoard grain.

How, people are still wondering, did England manage to avoid violent revolution? In the 1600s bread alone was consuming more than half of the labourer's family budget and the experts calculate that 80 percent of the workforce was involved in agricultural production; the remaining 20 percent was 'overhead' including nobles, merchants, artisans and clergy. It seems an impressive feat for such a relatively small group of stationary bandits to keep the hierarchy stable and the country subdued. They relied on public torture and executions, of course, but the inhumanities of that era pale into insignificance beside those of today.

When the poor could not afford to eat, there were often charitable arrangements for feeding them, and it has been estimated that half the working population had to rely on charity sooner or later. Since wages were set by law, not by the market, the frequent need for charity implies that wages were set low enough to prevent the workers from building up savings. The charitable relief was made easier to control by laws that prevented labourers from moving to another district, and so allowed micro-management on a local basis. Not only could they not get charity outside their district, but labourers were whipped or put in the stocks as vagabonds if found travelling without work. It was also illegal under the Statute of Inmates for one labouring family to allow another to share their dwelling. It was possible that a labourer's child could leave the land if it found a master who would take it as a servant or apprentice, but a 14th century law stipulated that if a son could not find an apprenticeship before the age of twelve he had to remain a farm labourer all his life.

This description of the planned economy before the seventeenth century economic revolution is a snapshot showing how tyrannical the traditional agriculture-based society had become in its obsession with preserving the social order and avoiding famine and riots.[42] To ensure stability monarchs had given away monopolies, land, and road tolls to their favourites to create a loyal hierarchy dedicated to preserving the social order. Compulsory tithes were paid to the church to fund priests who exerted thought-control from the pulpit. Any spare manpower in years of good harvest was diverted to cathedral-building. We can recognise the bountiful years from the spurts of growth in the cathedral towers. Farmers were forbidden to carry their produce out of their local area, local authorities synchronised planting and harvesting and decided how much grain was surplus to seed requirements and could be used, for example, to make beer. Roles and occupations in society were not learned, they were inherited. Workers who left their parish could not get poor-law relief elsewhere, and punishments for minor crimes were draconian.

If the tribes had known what they were letting their descendants in for, they would have fought Caesar to the death.

The heredity delusion

The secret of preserving this exploitative economy lay in a finely-honed cultural indoctrination. It is remarkable that no attempt seems to have been made until recent times to choose the apparently easier route of selectively breeding humans to suit hierarchy's needs. Humans had, after all, discovered very early the trick of selectively breeding wild animals to improve their meat, wool, strength, and docility. The reason for this omission appears to be that humans came to believe that they *had* been bred to occupy the roles allotted to them by tradition. Culture so successfully mimicked speciation that it became a universally held delusion that everyone's traditional station was inherited, having been engrained in the procreative seed by repetition of the same functions from generation to generation. No better evidence for the strength of this extraordinary delusion is needed than that Darwin believed it and put forward a 'scientific' theory for it – Pangenesis. He thought that acquired characteristics were passed on by a mechanism in which every cell in the body contributed its lifetime learned experience – and that of its ancestors – to the seed.

Recently in a discussion of how 'good breeding' was revered in Victorian times, someone suggested that the principle still holds today 'because a person's accent defines their class'. This statement is so *not* about breeding that it shows how counter-intuitive the whole idea of hereditary status now is – historians struggle even to get people to understand that such a delusion ever existed. Today your diction *does* say a lot about where you place yourself in the hierarchy, but 200 years ago it was less important because regional accents were common even among the aristocracy. The only thing people were interested in was who your father was and whether you were his oldest son. This information was of paramount importance, like the pedigree of a prize bull to a farmer at auction.

Of course, it was known that well-bred people were likely to be defective in intelligence, but intelligence was of no importance to 'station'. What counted was the qualification to *rule*, which had nothing to do with being able to add up or to understand philosophy or science. One could pay lesser people to do those things. The proclivity to rule over those of a lower station and be ruled by those of a higher one could only be acquired by passing down centuries of supposed innately accumulated experience in the fixed social order. The word *aristocracy* today still refers to breeding, but the last two syllables have now lost their meaning. They come from the Greek word for power, and that used to be the whole point of the word. This tends to be forgotten now that the aristocracy is just another group of celebrities.

In an England and Wales with only six million inhabitants in the mid-eighteenth century, the easiest way for workers to free themselves from the treadmill of agricultural labour and the consequent threat of starvation was to revert to hunting the abundant wildlife. The most revealing and ironic application of hereditary privilege was the system adopted to prevent this from happening. Under the Game Act of 1671 nobody could hunt wild animals unless he was the eldest son of a Knight or higher nobility or owned at least two or three hundred acres of high-quality farming land. This disqualified 99 percent of the population, including most proprietors of farms, and the privileged one percent took full advantage of their rights and their lack of any other occupation. Untitled owners of less than the required value of landholding not only had no right to hunt on their own land, but also had to let the privileged few hunt on it and trample the crops. Thus, in their favourite pursuit, the rulers could act as if the transition to boring, unhealthy settled agriculture had never happened.

What could be more agreeable for people with time on their hands and no education, who could feel that they were upholding the social order at the same time as reverting to a more natural state of humanity? There was no state police force at the time, and to enforce the law against poaching the aristocracy were entitled, and obliged by their sense of duty, to appoint gamekeepers with despotic powers. The gamekeepers had rights of arrest on suspicion, and resisting such arrest was punishable by death; they also had the right to search for and to confiscate guns. The law was therefore double-barrelled as far as maintaining the social order: it forced the populace to live by precarious agricultural labouring on the farms of the elite and also disarmed them, making rebellion difficult.

When we use the word 'class' now to refer to our modern social hierarchy it tends to be a rather fuzzy term which recalls such categories as rulers, bourgeois capitalists, and proletariat. No such broad generalisations were needed in the pre-industrial society. One man had authority over another man by virtue of what was then called 'degree', as in the text of the Game Act already referred to which specified that only the very wealthy or *'the son and heir apparent of an Esquire, or other person of higher degree'* could hunt game on his own or other people's land. *Esquire* meant, with a few exceptions, someone who would one day inherit the title of Knight and be referred to as Sir Something Somebody.

But the complete set of rulers included all those who did not have to work for a living because they owned land. The bottom level was the wealthy 'Gentleman'; above him the titles or immediate prospects of inheriting titles and land began with the next rank up – that of 'Esquire'. The size of the landowning elites was stabilised by another clever invention: the laws which regulated inheritance which were finally abolished only in 1925. All large estates were locked into a system, familiar to readers of Jane Austen or George Eliot, in which the land and the title could only be inherited by one male. The vagaries of childbirth and disease often cut the links in the chain of inheritance, but the Crown could be expected to repair this by allowing appointment of a more distant relative to the title in what was called 'patrilineal repair'. All state military or civilian appointments were allotted based on inherited precedence, both to take advantage of the supposed hereditary ability to rule and to nurture it for the benefit of future generations of leaders from the same family.

This delusion that acquired characteristics were passed down from generation to generation does not seem so farfetched when we remember that the human foetus is born prematurely so that it can finish its gestation 'in the womb of society' or perhaps better 'in the womb of its class culture'. In strict body weight terms, compared with other mammals, the human should be born after eleven months.

After its premature emergence at nine months, brain wiring and other developments take place in the cultural environment that in other mammals take place in the womb. The event of human 'birth' is a long-drawn-out cultural process. The infant does not actually become a complete human being until years have passed – and by that time it is a human being of a particular 'degree', just as if its social status had been coded into the genes of its parents.

The multilevel aristocracy, from gentleman up to monarch, was a beautifully engineered hierarchical system. It used the human capacity for delusion which appears when the knowledge-creation system is disabled. It was much more effective than anything that genetic natural selection could produce, because of its multiplicity of layers. This allowed optimal cascading of control rather than a simple dominance where one powerful animal controls the rest.

The power of the pulpit

The secret of this period of mind control was the farewell gift of the Roman Empire: state religion. Europe had become a theocracy as rigid as East Asian states today.

Before the climate stabilised, without hierarchy, without government, and without belief in the afterlife, our knowledge-sharing ancestors had tamed fire and invented painting, sculpture, toolmaking, medicine, cooking, clothing, and ocean navigation. Finally they had discovered agriculture and domestication of animals as soon as the weather made it possible. After the imposition of coercive religion, European humans invented nothing for the 900 years of the Dark Ages. Soon after the collapse of Roman authority in the European Reformation of the 1500s, they suddenly began inventing and innovating again.

Although the flame of independent thought burned in every breast throughout the Dark Ages, the Church hierarchy suppressed its diffusion by draconian measures. Those in authority were convinced that depriving individuals below them of their power to innovate would empower rulers to exercise more secure level of organisation. This, they thought, was needed to cope with the riots and famines that plagued their planned economy and which, as later developments showed, were the products of their rigid control.

The hierarchy had discovered humanity's weakness: natural selection's primitive communications protocol that makes humans accept repeated statements as true. Education, such as it was, was from the pulpit. The universities, far from being designed as centres of research, were founded to train clergymen who would provide the front-line defence against destabilising innovative thought.

The process of cultural dumbing-down had begun in earnest under the Roman Emperor Constantine who realised that a new charismatic religious movement could be of use to him in governing Rome's unruly Eastern Empire, and converted the Empire to the new variant of Judaism. Constantine's Council of Nicaea in 325 AD hammered out a creed that forbade independent thought by using mystical language that defied explanation or contradiction. The difficulty or impossibility of understanding how one god could be three at the same time was the key to the success of the Nicaean Creed. A person's professed 'faith' confirmed their rejection of logical thought.

Constantine's scheme took an unexpected turn after his death when the tax-exempt and therefore by then fabulously enriched Church took over what was left of the Roman Empire. But although Constantine's successor Justinian closed down Plato's Academy in Athens in a further attempt to eliminate logical thought, the Church had itself been fatally contaminated because the brand of Judaism which Constantine had co-opted included a radical component – Christianity – based on the teachings of a man who had more in common with Plato than with Moses. There were repeated official attempts to marginalise Jesus from the new religion in pursuit of a 'Judaism Lite' which would be acceptable to Gentiles because it did not require circumcision or dietary restrictions.[43] The solution was to merge Jesus into the Trinity and to deny that he had been human at all. But the mystic Galilean (possibly a closet agnostic himself, to judge by his disputatious asides) had implanted a spirit of dissent within the Church which undermined the whole principle of revealed religion. It would prove to be the Church's eventual undoing.

Christians who engage in debate with zoologists over the truth of Genesis have already lost the battle. Genesis is supposed to be the word of God, not a discussion document, and to argue about it with unbelievers is atheistic. Other revealed religions forbid this deviation, which may help to explain why they have lasted better.

Religion became all about conformity and obedience to hierarchy rather than spiritual belief. The old Christianity was first implanted in Britain by seasick charismatics stumbling ashore from rickety fishing boats. It proved to be so useful that the Roman Emperor Constantine's scheme of dumbing down his subjects through religion became a keystone in the architecture of the social order in Britain too. Spiritual matters came to take a distant second place behind the issue of religious obedience to secular authority which was much easier to understand and teach. A seventeenth century handbook for the instruction of the farming community says that the Fifth Commandment's instruction to '*Honour thy father and thy mother*' will be understood by all good citizens as a command to fear and respect all their betters including their '*landlord, or the gentleman his neighbour, because God hath placed them above him and he* [the good citizen] *hath learnt that by "the father he ought to honour" is meant all his superiors.*'[44]

The Church of England's catechism, which all children had to learn by heart, required the child to promise to '*submit myself to all my governors, teachers, spiritual pastors and masters; to order myself lowly and reverently to all my betters; … to learn to labour truly to get my own living, and to do my duty in that state of life unto which it shall please God to call me.*' Until the eve of the Industrial Revolution, those 'dissenters' who would not repeat this creed and therefore could not attend the Church of England were legally persecuted, and even until 1828 they still could not work for the state. Many of them (being generally and not surprisingly more successful and able to afford the fare) colonised America, leaving the rural population in Britain thoroughly cowed by a state religion that allowed no independent thought or hope of betterment.

The elitism of literacy

State religion was only one of the mechanisms which were effective in restraining the human drive for knowledge sharing. Possibly the most effective knowledge-limiter, paradoxically, was literacy. We think of literacy as the great liberator but its initial effects were the reverse. As late as the seventeenth century fewer than a third of men could even attempt to write their name, and fewer than ten percent of women. A far bigger majority of the working population was functionally illiterate, reading and writing being widely discouraged among the working class until the nineteenth century when industrialisation and empire-building began to require it. The inability to decipher written documents cut the population off from any new knowledge that humanity was creating, even the knowledge of new agricultural techniques, and made it easy for the literate to lord it over them by claiming exclusive familiarity with the classics, the newspapers, political tracts or any other knowledge repository.

The entire Stone Age population had been, of course, illiterate so how did literacy contribute to the dumbing-down that followed the transition from hunting to agriculture? The answer, as I have hinted, was the elitism that resulted from the interaction of literate and illiterate groups. This was an important component of the psychological power that the elite employed to deprive the people of thought, a power differential that did not exist in pre-literacy foraging days. The new knowledge enslaved the working classes who invested written material with magical qualities and knew that they were excluded from it. The only information to which they had access was endlessly repeated from the pulpit. How could they, with natural selection's outmoded repetition-coding input device, fail to believe it?

The mathematics of mass delusion

It may seem counterintuitive that hierarchy inhibits knowledge creation, because hierarchies seem to represent a universally superior structure that appears throughout the natural world. Without it, life would not have advanced beyond the pre-cellular 'organelle' stage. An organic hierarchy makes use of surplus energy created at the lowest level to feed higher levels. If these higher levels evolve to perform the function of governing the level below in such a way as to keep the surplus energy flowing upwards, the structure becomes miraculously stable. An example is our body's anti-cancer mechanism, which regulates and slows down the natural tendency of cells to reproduce themselves indefinitely.

Two mathematical approaches seem at first glance to be relevant to explaining hierarchy's success in creating mass delusions by repetition coding. One is Graph Theory and the other is Nobel Prizewinning economist Herbert Simon's analysis of the properties of hierarchies in his *Architecture of Complexity* (1962).

Mathematicians have used Graph Theory (the mathematics of objects connected together) to model the progress of a property that spreads among connected objects by travelling along the links between them like a contagion. The 'property' in the case under discussion would be information diffusing itself across a network of human brains.

Graph Theory seems to cater for a less than benevolent hierarchy. It defines a hierarchy as a subset of objects which are linked vertically in a pyramid of two or more layers. Theory holds that this hierarchical structure acts as a 'suppressor' of information flow because events appearing within it take a long time to infect the entire collection of objects by contagion. The problem in applying this to modern society is that this is not how a human hierarchy directs its communications. A human hierarchy such as a powerful democratic political party has a large media office which feeds to the news media an endless supply of free news about the successes of this leadership faction and the failings of that other leadership faction. Journalists on the lookout for low-cost news diffuse this in the first half of each day's news bulletin. This 'directed' connectivity, in Graph Theory terms, is a configuration that 'amplifies' the spread of contagions, not a 'suppressor' like a normal hierarchy. The constant refrain of leadership news makes both the population at large believe a repetition-induced illusory truth that human progress is dependent on hierarchy. They will even automatically believe that the intellectual deep freeze of the Dark Ages must have resulted from a collapse of Roman leadership when in reality the Roman hierarchy was just getting started. This focus on leadership has corrupted the whole of recorded history.

Turning to Herbert Simon and his *Architecture of Complexity*, it was Simon who invented the parable of the two watchmakers. One of them built watches in subassemblies and then assembled them in a hierarchy, while the other built the whole watch in a single assembly of discrete parts. The hierarchical approach proved more productive if one assumed that the latest partially-completed assembly fell apart during inevitable interruptions of other business.

Professor Philip Agre, of the University of California, criticised Simon's paper for implying that hierarchies are the always the most efficient way to organise complexity. Agre explained that Simon may have been over-reacting to the fashionable Systems Theory approach which placed too much emphasis on self-organising, non-hierarchical systems. According to Agre, Simon really believed that complex organisations used both hierarchy and self-organisation. For example a firm organised hierarchically would use spontaneous cross-departmental linkages to diffuse information. New communications technology, according to Agre, helps both hierarchical and non-hierarchical organisations equally.

This formula doesn't capture the properties of an organisation in which the self-organising component creates beneficial innovations that destabilise the hierarchy and the hierarchy responds by going on the offensive. This would be a more realistic picture of human society, where anarchism is always operating under cover. In industries such as research and development or management consulting which depend on brilliant ideas 'metastasising' (i.e. infecting every part of the organisation), hierarchy is the worst possible structure and is widely shunned. Neither does Agre's explanation consider new communications technology's facilitation of the mass delusions to which humans are vulnerable.

We should not despair at hierarchy's apparently successful mind control. Humans may have the most controlling hierarchical structure ever known, but it is not perfect. Like cancer, knowledge leaks through into the Cloud where it stays forever, as Galileo proved when he recanted under pressure his claim that Earth revolves around the sun. And as the next chapter shows, anarchic innovation robbed the state of a large portion of its power during the (partial) Enlightenment that followed the European Reformation

CHAPTER 6

Capitalism vs. Hierarchy

THE IMPACT OF SEVENTEENTH-CENTURY CAPITALISM on our culture has been misinterpreted both by its admirers and by its critics. Its admirers credit it with inventing innovation, ignoring the innovative explosion of the Stone Age prior to the clampdown of the hierarchical agricultural society. Capitalism's critics convict it of enslaving us and fostering inequality, ignoring the fact that inequality had reached its peak under the pre-capitalist regime of the previous 2000 years after a period of egalitarianism a hundred times longer.

Modern capitalism, in the form of the Industrial Revolution, began rather suddenly in seventeenth-century England. No modern economic historian acknowledges that a Stone Age over-productivity mentality or the innate individual desire for approval from the group had any role to play in this development. Strange as it may seem, the man who is widely credited with being the brains behind modern capitalism, Adam Smith, was convinced that both were critically important and that there was nothing new in capitalism.

Karl Marx, widely seen as capitalism's most vociferous critic, also factored human evolution into his historical theory. Marx's well-deserved bad reputation may explain why modern economists are resistant to an evolutionary approach. In this they do capitalism a disservice by giving credence to the idea that the culture of capitalism is entirely responsible for our inequality and environmental problems. It is our over-production, liberated from a few of the more restrictive practices of the state, but not from all state control, that are responsible for capitalism's successes and failures.

How capitalism reduced state power

Capitalism arose from a new Agricultural Revolution which preceded and facilitated the more famous Industrial Revolution. Step-changes in agriculture had begun to prove Malthus wrong even before he had expounded his theory that agricultural production could never keep pace with population growth. Joyce Appleby, in *The Relentless Revolution: A History of Capitalism* describes the causes and effects of the Agricultural Revolution. England by then benefited from later marriage age, privatisation (enclosure) of common lands, a relatively unified internal market, early limitation of the powers of the Monarch, a joined-up legal system, and relative publishing freedom. These led to circulation of new ideas for increasing agricultural productivity. This helped to reduce the labour force in agriculture from 80 percent in 1690 to 36 percent in 1800 and then to 25 percent in 1850. These anarchic agricultural improvements quickly eroded the long-established mindset that felt that any change to the established order was foolhardy. The breakthrough encouraged capitalists to use the labour freed from agriculture in more speculative ventures that led to the Industrial Revolution.

The traditional agricultural system had been a Soviet-style totalitarian planned economy. The discovery by eighteenth century economists that everyone was better off in a free agricultural marketplace weakened belief in the benefits of authoritarian government in general. But Appleby fails to justify her claim that the discovery was triggered by specific ideas relating to the benefits of capitalism. It resulted simply from the demonstration that the old belief in the benefits of a rigidly planned economy and the avoidance of change was mistaken. There were few capitalist ideas *per se*; the current of thought that swept England and then Europe simply questioned the validity of the old truths. No *system* of capitalism replaced the old system; instead, the economy reverted to a state of creative anarchy, with the difference that it was now highly taxed in ways that did not obstruct innovation. All that government had to do to implement capitalism was to stop monopolists and other abusers from manipulating the market and then wallow in the increased tax yield. Talk of the 'free market' as if it was an institution set up to optimise trade obscures Adam Smith's dictum that a free market exists within the breast of every individual. He claimed that our inclination to 'truck and barter' was innate.

The invisible Adam Smith

Adam Smith did not invent capitalism; he pointed out that it arose from innate human behaviour. Modern economic historians disagree with him and avoid quoting his views on the subject.

He was not a tea-chandler or a coffee house stock-market *habitué* as one might imagine from his popular reputation as the inventor of capitalism. He was an eccentric and absent-minded academic who earned his living as a tutor to rich aristocrats. He had the unusual and formative experience of being abducted as a toddler by a gypsy woman. The abduction lasted only a few hours until the woman, who had attracted attention because of the extreme distress of the infant in her arms, was chased by a mob and threw the screaming future political economist to the ground before escaping. A book remains to be written about how this incident and its constant retelling within his family may have unleashed Adam Smith's genius and inspired his psychological insights into human empathy.

Before Adam Smith became one of the most famous economic pundits of all time he was a pioneering psychologist. Surprisingly, he first used the now-famous term 'invisible hand' not in an economics treatise but in his revolutionary treatise on psychology, *The Theory of Moral Sentiments* (1759). By 'moral sentiments' Smith meant what we call emotions. This was a daring choice of subject; few philosophers in the religious culture of his day cared to probe the ways of the human mind. They found it safer to stick to visible externals or divine purposes to explain human behaviour. Smith, in his later and very famous book *The Wealth of Nations,* never used the expression 'invisible hand' to refer to a supposed self-regulating ability of a free market, although nowadays he is usually cited only in connection with that later concept.

In *Moral Sentiments,* which he always considered his best book, he used 'invisible hand' in a very particular sense – to describe an irrational innate drive within each of us which leads us to overproduce and in doing so unintentionally benefit our fellows. This drive, he maintained, acted by way of generating pleasure in a farmer, for example, by letting him consume his entire surplus, however mountainous, in his imagination. The unintended benefit to his fellows arises because there is nothing he can do after consuming as much as will satisfy his and his family's basic needs except to exchange it for symbolic goods which would not satisfy anyone's basic needs. Because the rich landowner has created and disposed of a surplus of something that satisfies basic needs, others who could not otherwise have satisfied their basic needs must consume it. This is also the principle behind Stone Age group cooperation and Smith makes a point of saying that its only reward, as then, is the acquisition of status symbols which advertise the creation of surplus by the landowner and his forebears.

In *The Wealth of Nations,* Smith again used the metaphor of an 'invisible hand' to describe a different innate human drive which has unintended consequences, showing that the phrase was a generic one describing any unintended consequence. He used it in the context of some merchants' desire to keep as much as possible of their capital assets near to themselves even when they employed them only in trading between distant countries. A risk-averse English merchant organising a trade in spices from China to Europe, wine from Lisbon to Jamaica, sugar from Jamaica to Hamburg, and herrings from Hamburg to Lisbon might be fearful of typhoons in China, slave revolts in the West Indies, plague in Hamburg, and embezzlement in Lisbon. To protect against these the more cautious merchant might build himself a nice brick warehouse on the Thames in London to store the merchandise, routing it through London on British ships. The merchant's desire to play the low-risk, low-return game, Smith said, would result in the enrichment of Britain, the country where he lived. This positive consequence for the nation's wealth would be unintended, i.e. be caused by an 'invisible hand' acting inside the merchant.

Smith's argument gives us an interesting picture of the worries that might give sleepless nights to a merchant in Georgian London. In his day this insight was a powerful argument against only one specific type of regulation: the government's mistaken attempt to increase national wealth by banning the export of money. It's not clear that in the modern era of global communications and relative political stability the particular 'invisible hand' that Smith was describing would have any power to prevent City of London financial speculators from undermining the national economy. Nevertheless modern thought has attributed to Smith the peculiar idea that an unregulated market is best for innovation and trade.

Smith only used the "invisible hand" metaphor in one other unique sense, describing the gravitational attraction between planets in a book about astronomy. (The book, because it described an astronomical system independent of a Creator, could not be published in his lifetime). 'Invisible hand' was therefore evidently a neutral phrase that he used three times to describe different tendencies outside intelligent control, and he never connected it to any market. Despite this, economic history has invented a supposed market mechanism called 'Adam Smith's invisible hand of the market.' The complete phrase thus falsely attributed to him was not even used until the golden age of *laissez faire* a century after Smith's death.[45] Of course, the fact that Smith knew nothing about it and would have opposed it tooth and nail does not detract from the value of the idea, which may have a much more venerable provenance. It may be simply the *laissez faire* merchant's version of the ancient doctrine of the Divine Right of Kings.

Adam Smith, now widely touted as an apostle of *laissez-faire* and opponent of market regulation, only ever used the shorter phrase 'invisible hand' to criticise one mistaken type of regulation. This was the law that prohibited the export of money and handicapped the risk-averse merchant, described above, who needed to pay his suppliers around the world. Elsewhere in *The Wealth of Nations* Smith called explicitly for more regulation of markets, not *laissez faire*. For example he predicted that a loan default crisis would ensue from the government's failure to regulate banks.[46]

Smith believed, correctly as we now find, that capitalism was innate in humans. Even a forager would have a natural predisposition to 'better his condition' and to 'truck and barter' because arrow-making, (Smith's example), required specialised skills and not everyone would learn them. (Modern economists prefer his 'pin factory' example which avoids the Stone Age angle). Smith's psychological vision is of a human with an innate tendency to create surplus because of the gift of being able to get satisfaction from advertising his value to society. This psychology has recently been detected in early humans, as described in previous chapters.

Smith is the strongest proponent to date of the idea that humans have an innate predisposition to 'better their condition', but one must turn to his *Theory of Moral Sentiments* to find that this does not simply mean individuals increasing their consumption or amassing bullion. It includes increasing their pleasure in less obvious psychological ways, notably by providing what Kropotkin called mutual aid.

Although *Moral Sentiments* was first published seventeen years before *Wealth of Nations*, the last act of Smith's life was to revise it and the two books are compatible and complementary, one describing the national interest and the other the individual's interest. He was firmly on the side of innate cooperation, rather than any supposedly new capitalist ideas, as the ultimate cause of economic growth.

Naming an Institute after Adam Smith that lobbies against all kinds of government regulation of commerce is an example of a very successful repetition-induced illusory truth.

The divisive Karl Marx

Karl Marx, the leading critic of capitalism, was born in 1818 to a Jewish family who had converted to Christianity to avoid discrimination in their native Germany. He grew up in the feudal agricultural society of Prussia. There he nurtured radical ideas that he transferred to the Industrial Revolution after he fled to Britain.

Marx did have some insight into the tyranny imposed by hierarchical organisation. He simplified the concept by portraying it as an oligopoly – a two-level hierarchy where a small minority of exploiters (the capitalists) dominate the rest. In his view the capitalists had amassed their wealth by piracy, slave-trading, robbery, and other parasitical economic ventures and then had 'gone legitimate' like a Mafia family by investing the proceeds in the new more productive industries. His theory was supported by the fact that in England, the cradle of the Industrial Revolution, many wealthy families had been at the forefront of the world slave trade. Others had stolen Spanish gold from the Americas through 'privateering' in which their armed civilian ships were licensed to raid the Spanish colonies under the pretext of war.

These thieves, Marx claimed, were unproductive in the new industries. They were able to extract the wealth because they were the only group rich enough to buy the labour which was the source of the new industrial productivity. The manual labour was provided by the working people who had been driven away from subsistence farming by enclosure ('privatisation') of common land. It was a plausible narrative: Britain, the nation most adept at colonising, robbing, and enslaving foreigners had turned to colonising its own peasantry, stealing their land and enslaving them in factories.

This binary unproductive/productive class model was better suited to the feudal agricultural systems of Marx's childhood in Prussia, in which the landowners took no part in production at all. It ignored the fact that industrialisation was different in that manual labour was only a part of the value-adding process. Research and development, organisation, and distribution were a new requirement and British capitalists became involved in these areas, funding them through reinvestment instead of squandering profits on conspicuous consumption as in the old agrarian system. They were not a truly unproductive group who could be dispensed with, so the historical class struggle idea (in which in Marx's view either one of the classes would be inevitably destroyed by the 'dialectic' between them) was not as good a fit as it seemed.

Marx's 'us versus them' model was acceptable to those who advocated violent revolution and the 'dictatorship of the proletariat'. The 'exploiter' level of the two-level oligopoly was an easy target for violent overthrow and replacement. A simplistic view that a replacement Marxist top level would be dedicated to advancing the interests of the lower level solved the problem that the proletariat was working too hard to govern itself. All that was needed, under Marx's authoritarian communism, was to replace the top level with an elite group of communists dedicated to interpreting the will of the proletariat.

Marx and his followers did not understand or acknowledge automatic hierarchical behaviour, in which each upper level of a hierarchy develops goals of its own, primarily ensuring its own stability at any cost. This 'iron law of oligopoly' (as sociologist Robert Michels called it in 1910) was recognised by other radical leaders including Kropotkin and caused a profound split between authoritarian and liberal socialist parties in Europe.[47] Unsurprisingly the authoritarian Marxists won the day; they had the laws of physics on their side in the shape of the universal law of hierarchy. This law is subtler than the 'iron law of oligopoly'. Two-level 'oligopolies' can exist, but the naturally occurring form has multiple levels in which each level aspires to promotion by showing that it can command the loyalty of the level below. The word 'hierarchy' derives from 'sacred rule' – the system that ranks the angels in multiple levels of holiness.

Marx had studied the anthropology of the day and agreed with Adam Smith, and with this book, that economic growth results from humankind's innate faculties, not from new capitalist ideas. But if he believed that Stone Age humans had been communists in the sense of holding wealth and the means of production in common, he was mistaken. There was no wealth, and the means of production was the skill possessed by individual hunter-gathers that led them to 'truck and barter' in good capitalist style.

The ethical Max Weber

The pioneering sociologist Max Weber, who tried to undo the damage caused by Marxism, was born in 1864, when Marx was 46. He died in 1920, a victim of the influenza pandemic that killed more people than the First World War. He did not agree with Marx and Smith that capitalism or the incentive to work was innate in humans. Modern economic historians tend to follow Weber. This may be because they have not examined the latest evidence from the Stone Age.

Max Weber's name is not nearly as widely known as his catch-phrase 'The Protestant Ethic' (the title of his most famous book) or 'the Protestant work ethic' as it is more popularly known. As a sound-bite the title triggers some interesting thoughts. Most people will conclude that it must be referring to something non-religious otherwise the two words would be saying the same thing, as in 'Protestant belief'. Furthermore, it sounds as if it refers to something praiseworthy. 'Nazi work ethic' might not strictly be an oxymoron, but it would be a bizarre concept.

Having listened to the context to find out more about Weber's phrase you would discover that it refers to the supposedly increased motivation of inhabitants of traditionally Protestant countries to get out of bed in the morning, with positive results for the economy. The phrase implies that agnostics or atheists raised in Protestant cultures can realise the ethical benefits of religion through work without having to spend time in church. It is not a phrase that rolls so easily off the tongue in traditionally Catholic countries; the French are more inclined to speak of 'Anglo-Saxon ideas', which does not carry such an admiring subtext.

Weber, like Marx, was a workaholic and a believer in history's ability to explain progress and industrialisation. He was in the mainstream of society, being a respected member of the academic establishment rather than a poverty-stricken revolutionary pamphleteer like Marx. He was keen to rebut Marx's disturbing theory that progress had resulted and would continue to result from violent class struggle, although he accepted that modern capitalists

72

were often immoral. But where Marx attributed historical developments to the economic interests of different classes, Weber wanted to show that they resulted from religious conviction.

The biggest difference between Weber and both Marx and Smith, however, was that Weber believed that no human had any innate motivation to overproduce. "A man does not by nature wish to earn more and more money," wrote Weber, "but simply to live as he is accustomed to live and to earn as much money as is necessary to earn for that purpose".

While Marx believed that increasing production was an innate desire of humankind, and that this quality had been exploited by a class which had seized control of wealth, Weber believed that cultural pressures made people work hard, pressures which represented the higher aspirations of the group. His explanation reduced conflict between worker and capitalist, so it was reassuring to those worried by the social tensions encouraged by Marx. On the basic disagreement over whether economic progress came from individual human nature Weber was also in opposition to Adam Smith who believed that mankind had an innate tendency to overproduce. So of the three of them Weber was the only one in favour of what we might call 'zero percent nature, one hundred percent nurture' hypothesis for explaining over-productivity and economic forces.

In Weber's view of how 'society' persuaded humankind to over-produce, industrial progress and capitalism resulted from the decentralisation of power. The first step was always the separation of church and state, and the next was appropriation of some of the monarch's power by the barons, creating a feudal system. Then, cities became autonomous and appropriated some power from monarch and barons. According to Weber, professional guilds and other local power groups in these autonomous cities kick-started the Industrial Revolution. Centralised power hierarchies, he realised, were never in favour of innovation or industrialisation because these empowered the masses and thus disturbed the hierarchical social order. This part of Webber's analysis has much in common with Kropotkin's belief in the harmful historical role of centralised power.

Weber believed that for industrialisation to succeed in a decentralised state, both capitalist and worker would need to have the same religion. The religion must be classless, which ruled out Hinduism but suited Christianity. It must be intolerant of other religions and have worldwide ambitions. This also suited Christianity which had no compunction about forcing conversion, by the sword if necessary. Christianity had converted under duress many of the Jewish Diaspora as well as colonial victims. Islam, though an expansionist religion like Christianity, was more tolerant. It did not insist on the conversion of its conquests partly because its economic system depended on levying higher taxes on unbelievers.

Weber also believed that an industrialist religion had to be ascetic and not too fond of conspicuous consumption or charitable giving, so that further investment to increase production would be a normal way to dispose of profits. All this conveniently pointed to Protestant Britain coming out on top as the place where the Industrial Revolution was bound to start.

Weber's greatest handicap was that he did not know how successfully hunter-gatherers had adapted to frequent change in the distant past when he proposed that a new social ideology – capitalism – had been needed to haul humanity out of its (he thought) timeless agricultural rut.

Unlike Adam Smith and Karl Marx, Weber's name is unknown to most people, but his catchphrase "the Protestant work ethic" still brings comfort to those who have abandoned the religion of their parents but fancy they retain the ethics by braving the rush hour at the beginning and end of each day and office politics in between.

The new Weberians

Economic historians now treat capitalism as a purely cultural phenomenon, as Weber did. They don't describe or explain any innate human propensity (other than 'greed') like the more positive motivations that Adam Smith and Karl Marx believed in, and they do not explore the heritage of human evolution. The study of capitalism today seems to accept the Weberian view of it as a system recently imposed by elites from the top down rather than something built into the human phenotype.

The conventional economic wisdom, as articulated by Joyce Appleby, rejects the theories of Marx and Smith on the grounds that those two thinkers 'gave attitudes to men and women that they couldn't possibly have had before capitalist practices arrived'. Appleby argues that only the arrival of 17th century capitalism can explain 'how the values, habits, and modes of reasoning that were essential to progressive economic advance ever rooted themselves in the soil of premodern Europe.' But Marx and Smith had found these values rooted in an older garden, having studied what was available from anthropology.

In giving all the credit to seventeenth century economists' rebuttal of existing belief in a planned economy, modern economists pass over the fact that society had been free of that belief for many millennia before the introduction of agriculture and had benefited from a free market during that time. They use the phrase 'traditional society' to refer to *farming* society rather than the pre-farming human era of the Stone Age which lasted a thousand times as long. For today's economic historians, elitism and greed are the great industrial enablers, and humanity's evolutionary trajectory is not part of the picture.[48]

Advertising, not greed

Rather than instilling greed in the populace, seventeenth-century capitalism's contribution to prosperity was innovative symbolic signalling of the group's approval of individual effort. This mobilised the latent desire to show off the fact that one's efforts met with group approval . Adam Smith explains how the signals work for the rich landowner who, with his admirers, is deluded (Smith's word) that the utility value of the luxury goods he can obtain in exchange for his surplus grain is equal to their huge symbolic value.

The fine stately home provides the rich landowner with what Smith derides as 'a few trifling conveniences to the body' at vast cost. His clocks, made by master craftsmen, are accurate to within a minute a day in case he wants to be more punctual than anyone he is ever likely to have an appointment with. His sporting guns have internal components polished to mirrors so that they will last a hundred lifetimes without rusting. He has ice-houses to startle his visitors, redundant carriages of all sizes, Old Master paintings on which to rest the eye for a few moments each day, antique statues, and so on. These are 'positional goods' that have no real value to life but which are symbols of the cumulative size of the rich family's creation of surplus subsistence goods over the years, not affordable by anyone producing less.

The 'different baubles and trinkets, which are employed in the economy of greatness' (Smith again) have been designed to be as expensive as possible, not as useful as possible. These useless but highly symbolic 'positional' (status) goods are all that the rich landowner can possibly get in this world in exchange for that mountain of life-saving wheat. They are permanent reminders to his admirers of what he and they thought was greed but was the invisible hand of involuntary mutual aid. Without this need for display, our cooperative physiology and behaviour could never have evolved in the first place.

Before capitalism arrived, the rest of us tiny cogs in the production machine could not flaunt that kind of symbol of the group's approval. All we had to show off after a day's work was a few columns of figures we had written in an account book, an empty shelf in the shop where we worked as a sales assistant, or some dirty tankards in a pub where we served behind the bar. This problem is one that capitalism solved in the seventeenth century by discovering the greatest productivity aid in the history of humanity: *the silk waistcoat.*

As Appleby tells it, ships of the London-based East India Company began arriving in London with brightly-coloured textiles made in their eastern workshops, and the country was entranced. Within a few years the rich were astonished by the appearance of flamboyantly-dressed nobodies in Britain's streets, the young men shooting their cuffs and smoothing their brocaded waistcoats while their female companions swished their bright gingham dresses, all made available through the efforts of the East India Company.[49] The free market had by accident discovered *how to make positional goods for the poor.*

The sight of the lowest rung of society flaunting evidence of their labours revealed to the elite the astonishing fact that ordinary people would work harder than necessary if they were given the means of showing off symbolically the fact that they had created a surplus for society. Those sluggard paupers would not do a stroke of work more than necessary to keep themselves alive when the only commodities allowed to them were bread and ale. Now they were as easily deluded as the rich landowner when offered the opportunity to flaunt symbolic tokens of the surplus they created for others.

It did not take the capitalist societies long to exploit the profitability that resulted from every layer of the hierarchy flaunting symbolic success. The more liberalised societies roared ahead economically, creating ever more positional goods enabling individuals to signal their contribution to mutual aid.

Non-hierarchical capitalism

Although the innovation-inhibiting effect of hierarchical organisation is a drawback, it is hard to implement an efficient non-hierarchical system of industrial production. This is as true for services such as health care as well as for goods. A compromise relies on outsourcing innovation, when required, to specialist firms with a non-hierarchical internal organisations. There is an analogous situation in the insect world: when intensely hierarchical honeybees need to seek a new hive, they temporarily morph into a non-hierarchical form.[50]

The most successful of the world's strategic management consulting firm are deliberately structured in a non-hierarchical way. I have worked for two and the contrast between them is instructive. The first firm had branched out into management consultancy from its roots in technology, and its top management remained rather fond of hierarchy. At a certain point the CEO launched a new global strategy for the firm. The heads of consulting in different overseas markets disagreed, so the CEO fired them and appointed younger people who would follow his vision. That was the moment when I quit after twelve happy years. A short time later that firm had to file for protection from its creditors when the new vision failed, and now only the name survives.

My next employer had been a pure management consultancy since its birth in Chicago more than seventy years before as part of the McKinsey firm. The internal culture could not have been more different from the Boston Brahmin culture of my former employer. The founder in Chicago had lived by the aphorism "the wisdom of the group". Dissent was welcomed and criticism accepted without rancour. Future strategy was agreed in 'town hall' assemblies of the partners. I was shocked to find that in project meetings it was compulsory for the partner in charge to invite a partner from another discipline so that team members could express their disagreements openly in front of an independent senior staff member with whom they had no hierarchical relationship. This firm continues to go from strength to strength.

The moral is that we may not be able to abolish hierarchical organisation, but should restrict it to activities where it does not conflict with innovation, such as in steady-state industrial production. Change and innovation should be delegated to non-hierarchical groups. This is how our hierarchical social system has survived with the help of anarchism, as the Industrial Revolution shows.

The area most in need of anarchist anti-hierarchical thinking, given the increasing prevalence of mass delusions, is our educational system as the next chapter will show.

CHAPTER 7

Daring to be Wise

M Y FIRST TASK AS A MANAGEMENT CONSULTANT was to attend boot camp to be taught verbal reasoning. It was one of those brain-changing courses that makes one look at the simplest day-to-day problems differently. For a while, even a choice between taxi and bus required a formal issue analysis. The course was taught by Barbara Minto,[51] who had been McKinsey's first female consultant. As far as I could see, she was training all the world's young business gurus. I had only an undergraduate degree but the other students had PhDs or MBAs from the best schools. They were as puzzled as I was as to why we had not been taught verbal reasoning in school or at university. It's not rocket science, after all. Must this material be reserved for an elite, just so that they can charge thousands of dollars a day for outthinking their clients?

My case was more striking as I had been rejected as a candidate management consultant before because I had badly failed a verbal reasoning test. Now I had been hurriedly recruited by a top firm which skipped the test because they needed an expert who could explain the internet technology to senior executives in three European languages. Twelve years later I was tested again for verbal reasoning by potential employers and found to be in the top few percentiles. On-the-job training is enough, it seems. We all have it in us.

We, Galileo

We must all be potential business gurus or scientists like Galileo. Otherwise what is all that unused brain doing inside our heads? If archaeologist Tim Taylor's '200 mph' brain analogy is a guide then 60 percent of our brain is 'junk brain', surplus to current requirements. And this is high-maintenance junk. You don't just turn it off like a light bulb when you're not using it and save the energy; it is burning up calories all the time. Junk brain by this estimate consumes more than a tenth of all the calories we take in, and a third of every human baby's food intake goes into building it for no purpose at all. If we could de-evolve this unused high-carbon junk brain, our metabolism could put the energy into something more useful such as repairing body cells to make us live longer.

Early humans used all of their brain to work out how to conquer the most terrifying dangers and turn them to advantage. They learned how to suppress the fear of natural fire caused by lightning or volcanoes and harness it to drive a predator away from its kill. Somehow, they learned to manufacture *curare*, a tranquilliser used in hunting which can only be produced by an intricate multistage chemical process.[52] The Early Stone Age was a time when only obsessive mental effort would do. This mental capacity is surplus to requirements in the current stable climate; there is no longer any need for any of us to study nature to survive. But until evolution can find a way of decommissioning the junk brain in every one of us, it will betray its presence by exploding sporadically and unpredictably in what we call 'talent'.

David Shenk, in *The Genius in All of Us* did not explore our Stone Age heritage. He could have strengthened his claim that everyone is a potentially talented by pointing out that our brains evolved to cope with a much more demanding environment in which extreme mental effort was compulsory. Humanity's newly-discovered 'deep history' neatly fills some gaps in Shenk's argument.

First, Shenk asks how can we have hidden the fact that everyone is potentially extremely talented? Why do we believe so fervently the myth of variable innate talents? The answer must our old enemy, hierarchy. Hierarchy has few job openings for the talented, most of them in 'winner takes all' forms of sport and entertainment where rarity is part of the attraction. A widespread delusion that we are born with different levels of innate talent helps people to accept their positions in the production pyramid. New 'evidence' of the innate nature of talents is continually discovered to confirm something that we think we 'know'. For example, the discovery that better myelin sheathing of nerves increases intelligence is presented as evidence that differences in intelligence are related to physiological quantities and are therefore 'inheritable'.[53] Such 'evidence' ignores earlier discoveries that intensive mental activity increases myelination, supporting the research highlighted by Shenk showing that extreme talent only appears after its owner has performed obsessive mental labour over a considerable period.[54]

Second, Shenk shows that developing a talent requires motivation, and the motivating agent can sometimes be identified, but he wonders why it doesn't fit any consistent pattern.[55] The spur can be early failure or early success, an encouraging parent or a discouraging parent, or any number of other triggers. Surely the reason is that our mediocrity is *unstable* because it is imposed by an outside force – hierarchy and the repetition-induced illusory truth effect. *Any* psychological disturbance could break the chains and release the potential.

Schooling for the hierarchy

In Britain today state and private education collude with each other to serve and maintain the hierarchy by regulating knowledge creation to create separate castes. The state system strives for inclusiveness while the private system has its eye on elite roles like the judiciary in which a more pedantic knowledge of grammar is necessary.

One Appeal Court judge when making a ruling in 2004 found it necessary to write 'It seems to me that, in terms of syntax, the plainly natural effect of "the same" is to refer back to the composite noun clause .... [56] That judge had been educated at London's most exclusive private school. Not many people educated in state schools would be able to formulate such a ruling; a study by Williams and Hardman in 1995 showed that three quarters of university-educated state school teachers didn't know the difference between a clause and a phrase. In 2014, not surprisingly, 71 per cent of senior judges come from the 7 per cent of the population educated at private schools.[57]

The better state schools (accepting about 2 percent of children) are 'selective', choosing pupils based on examinations that resemble intelligence tests. Birmingham City Council, the local authority responsible for education in Britain's second largest city, has a policy of not teaching certain subjects to primary school pupils because it would interfere with the selection tests. Its brochure for parents tries to explain why:

> The Local Authority's policy is that no teaching or coaching for selective tests should take place in any of the city's primary schools. It is believed that the practice of coaching is counterproductive, as overachievement in the selective tests may be against the long term interests of children whose ability may be more limited than their test score would suggest.

Birmingham's brochure for parents categorises the selective tests under four headings: verbal ability, non-verbal ability, reading comprehension, and numerical ability. An official 'familiarisation' brochure gives examples of questions that test what Birmingham educators call 'verbal ability' and most people call 'verbal reasoning': [58]

> "Mother's Day probably has its origins in Greek or Roman times. In more recent centuries, it has marked occasions when servants were granted an afternoon off work to visit their mothers. A commercial element has added a contemporary twist to the tradition with the advent of Mother's Day cards."

> Question 1: When does the passage suggest that the tradition of Mother's Day began?

a. No-one can guess when it may have started.

b. It is a modern invention.

c. It is likely to date back to Ancient Rome or Greece.

d. Last century, when servants were given time off work to see their mothers.

Question 2: What does the author say about the effect of Mother's Day cards on this tradition?

a They provide a money-making opportunity.

b They have brightened up the celebrations.

c They have twisted the real meaning of Mother's Day.

d They are not necessary for this historic celebration.

Birmingham's policy of advising teachers and parents not to teach verbal reasoning will ensure that most children will fail to develop the basic social skill needed to identify this kind of error. The policy itself provides interesting material for a verbal reasoning test. You may want to re-read Birmingham's advice to parents carefully and then, using only the information given in it, say which of the following statements is a reasonable conclusion to draw from the passage:

a. Selective tests measure a child's verbal reasoning ability;
b. Selective tests do *not* measure a child's verbal reasoning ability;
c. Both of the above.

This policy is clearly a test for discriminating among parents, not among children. Thoughtful parents would immediately deduce from it that they should privately educate their children or at least coach them for entry to a selective state school. A parent who was not able to see the biased logic would be reassured by the subtext that too much education is bad for children. Thus the appropriate proportions of elite and non-elite will reproduce themselves with uncanny precision just as if a heavenly power was appointing them to those roles or as if it was coded in their genes.

Why should a state dedicated to 'it's the economy, stupid', US President Bill Clinton's apt description of the focus of today's hierarchical government, do any more than is economically justified in the way of education? It is hard to find any grounds to criticise Birmingham's pragmatic way of staffing the different levels of a hierarchy.

I have picked on Britain in these examples, which would not necessarily apply elsewhere. Many people regard Britain as untypical because it has the lowest social mobility in the developed world.[59] It is not surprising that the country that was most rigidly hierarchical before the Industrial Revolution should be the most adept at perpetuating the system. It is only 12 generations ago that a 20-year-old anthropologist *avant la lettre* was hanged for blasphemy in Britain for describing the Old Testament as 'Ezra's fables' in a private conversation. The US was peopled largely by emigrants who were fleeing such regimes so one would not expect to find such blatant relics of the legacy.

The invisible hand of hierarchy

Rather than railing against government functionaries for deliberately stunting most children's development, it's better to think of them as involuntary tools of hierarchy. Hierarchy is one of nature's favourite structure, as Herbert Simon seemed to believe. When early humans rejected hierarchy emphatically two million years ago by choosing egalitarian mating arrangements, it was as if the bathwater decided it wanted to go down the plughole in a straight line instead of around in circles as ordained by the laws of physics. The universe has not forgiven us.

Hierarchy is not created by well- or ill-intentioned people. No fiendishly clever intelligence is responsible for crafting Birmingham's disguised parent-selection test quoted above. The functionary who drafted it presumably did not get coaching in verbal reasoning, and mistakes tend not to be corrected at a higher level unless they undermine the hierarchy.

would be useful to hierarchy if Earth's climate changed to allow agriculture!

Of course, hierarchy ensures different life outcomes based on trivial differences when it pseudo-scientifically but arbitrarily selects individuals for its unequal privileges. That does not affect the scientific fact that the human cooperative toolkit still exists, having evolved purely to ensure that all humans are created equal. Even society's most dangerous threat — the repetition-induced truth effect which makes all humans vulnerable to extremist robot propaganda and hierarchy's delusions alike — is shared equally among us all.

CHAPTER 8

Democracy vs. Hierarchy

THE PRECEDING CHAPTERS HAVE SHOWN how hierarchy held back historic human economic activity and how it is damaging our modern educational system. We will turn now to how it damages our democratic government.

Those who were in favour of giving the vote to working people in the eighteenth and nineteenth centuries worried that the people might adopt the worst habits of the old monarchies. They feared that a party or faction could take over a democratic assembly by abusing the electoral process. Time has shown that their fears were well-founded.

The spectre of majority rule

The granting of the vote to working people started in America with the Constitutional Congress of 1787 and in Britain with the Representation of the People Act of 1867. In both cases, the system chosen was 'representative government'. This is an electoral system where citizens vote to elect people to represent their interests for a limited time. Those elected meet to debate and make laws on behalf of the whole community. They knew that, unlike the time-rich landowners who alone had the vote up until then, a new working-class electorate would not have the time to study and debate issues and solutions.

In Britain's parliamentary system today, voters are no longer free to use their vote as the proponents of this system intended. Instead of using elections to appoint a representative assembly which will study the issues and make the laws, they are offered a choice between national hierarchies ('parties') offering branded packages of manifesto promises, values and dogma. These are produced by negotiation within each political party behind closed doors.

This would disappoint the politicians who pioneered the new more inclusive government, including John Adams and James Madison in America and John Stuart Mill in Britain. They wanted Congress and Parliament to be a 'representative assembly' which would bring the experience of all members of society to bear – the 'wisdom of the group'. They did not intend the rest of the electorate ever to have to listen to political speeches or to understand the finances of the nation. To understand the failure of representative government today, we need to understand what a 'representative assembly' meant when the term was first used.

In one of the most influential documents recommending the ratification of the US Constitution James Madison, later the fourth President, identified 'the various and unequal distribution of property' as the most important danger facing the new republic. He didn't know that prehistoric climate change had derailed our egalitarian society by creating property, but he clearly understood that differences in property and wealth divided people. Both he and the later British politician and philosopher John Stuart Mill knew that a minority group of property owners had long been oppressing a larger group of workers and tradespeople. They feared that the non-propertied voters, now to become a majority, could take over legislative assembly and pass laws for 'an abolition of debts, for an equal division of property, or for any other improper or wicked project', as Madison put it.[62]

Their proposed solution to this threat was to ensure that the assembly was as 'representative' as possible — a sample of the citizenry which contained many overlapping interests besides the 'haves' and the 'have-nots'. John Adams, later to become the second President, wished for an 'exact portrait in miniature' of the population at large. Another member of the Constitutional Congress described the ideal assembly as 'the most exact transcript of the whole Society'. The full-time members would have time to discuss and compromise between conflicting interests.

This was the original meaning of 'representative': a faithful sample representing the composition of the whole electorate and all its various special interests. In today's politics, 'representative' has come to mean something very different. A modern 'representative of the people' is an individual specially qualified in electioneering and explaining the party manifesto to voters. The selection of representatives does not nowadays replicate the composition of the population. Some party selection procedures do try to enforce ethnic and gender quotas based on the population at large. Those quotas do not conflict with the selectors' overriding goals of finding representatives who are loyal to the party and qualified in the art of getting elected. There are no quotas ensuring the election of people with the practical experience that keep things running in the outside world.

But in the 1860s, when the vote was given to British working people, it was hoped that these political parties or 'factions', until then seen as a minor influence, would be swamped by other interests in the larger and more representative assembly. Nobody could foresee how future mass-media technologies would encourage what Madison called 'the vicious arts by which elections are too often carried'.

Initial success

The British Representation of the People Act of 1867 (also known as the Second Reform Act) allowed a property-less majority to vote. Even though women and many working men were still excluded, it was like the end of apartheid. Working man and tradesmen would now elect most MPs and the 'landed interest' would no longer control the House of Commons.[63] The hierarchical power of the landowner would be eliminated, and the mixture of interests in

Parliament would prevent the new largest interest, those who did not own property, from having everything their own way.

The removal of landowner dominance led to a burst of social reform. The first general election in which non-landowners voted came in the following year, and the government that came to power lasted until 1874. It was called a 'Liberal' government but the use of the capital letter did not then signify that it was a single organised party; it meant only a government formed of progressive ministers. Members of the new Cabinet were often referred to as Liberal Whig, Liberal Radical, and Liberal Conservative (as the new Prime Minister, Gladstone, was officially designated at the time). Gladstone chose his cabinet to be a representative sample of all shades of 'progressive' thought, from progressive landed aristocrats (the majority) to Radicals.

The word *liberal* had until then meant economically liberal in the sense of not supporting high taxes and public expenditure. In contrast, *conservative* governments had traditionally levied high taxes on the less well-off workers (tariffs on imported wheat, and taxes on paper which also impeded the spread of seditious ideas). They had spent public funds mostly for the benefit of the rich (financing foreign wars, the established Church, and imperial expansion based on slavery and colonial exploitation).

The new progressive (or 'left'-) liberal government, opposed to taxing the poor to support the rich, instead chose to tax rich property owners to support the poor. These taxes could be most efficiently levied by local government, so Britain witnessed the unusual spectacle of a national government voluntarily giving away some of its power to Local Authorities.

The tax that Parliament chose was a recurring annual real estate tax ("rates") based on regular revaluation of each property. A Local Government Board Act in 1871 gave to municipalities many of the tax-raising, borrowing, and spending powers previously monopolised by central government. Local government now received and spent the rates, thus acquiring the financial ability to improve the public realm and public services. Improvements were self-financing by the increased tax yield from those properties that increased in value as a result.

The new non-landowning voters elected non-landowners to local government. These (often local businessmen) invested property tax receipts in publicly-owned utilities providing piped water, gas, and drainage. The innovative energy and bureaucratic pride of more than 400 local authorities allowed them to launch a financially sustainable welfare state. They quickly implemented local-government-funded secular schools, sanitation, food standards regulation, gas and water utilities, environmental protection (under the Alkali Act) and slum clearance.

Another forgotten achievement of the first parliament of the working class was votes for independent women if they had the same qualifications as newly enfranchised males. Yes, one year later, in 1869, Parliament enfranchised these women in local government elections. Today, with local government reduced to powerless 'branch offices of Whitehall', the importance of this step has been underrated and all the attention has been given to parliamentary enfranchisement in 1918 instead of to the first women's votes in 1869. But in 1869 local government was collectively more powerful than Parliament, because of its new tax-raising powers and autonomy. Voting was not yet secret, so 'dependant' women also had a chance to influence the parliamentary vote of the male head of their household. As a result of neglect of the female vote of 1869, women are implicitly being denied their role in launching the welfare state.

Prime Minister Gladstone had no means of rewarding loyalty by patronage. Previously, Civil Service employment had been the usual way of awarding sinecures to the family members of loyal followers or idle aristocrats. But Gladstone had committed himself to converting the Civil Service to competitive entry on merit. Apart from a few paid ministerial posts, he only had honours to give, and even these he tended to share on a non-partisan basis.[64]

Deprived of both carrot and stick for promoting party loyalty, Parliament had become a deliberative representative assembly capable of finding creative compromises between opposing interests. With a Cabinet chosen to represent all the shades of thought other than the reactionary, this first Gladstone government is widely considered to have been one of the most effective ever to hold office in Britain.

Oligarchy takes over

But we now know that nature abhors a vacuum where there is wealth and privilege to sustain a hierarchy. A new hierarchy soon sprang up based on the political party, which until then had been a very loosely organised structure. This spontaneous development conforms to this book's thesis that hierarchy behaves like an autonomous living entity.

It wasn't until 1910 that sociologist Robert Michels, formulated his 'Iron Law of Oligarchy' to explain why socialist politicians had failed to represent the workers who had put them in power. He was cynical about the attempts of the founders of modern democracy to guard against hierarchical rule, writing "Historical evolution mocks all the prophylactic measures that have been adopted for the prevention of oligarchy." Michels recognised that the hierarchical vacuum left by the formerly dominant aristocratic interests had been quickly filled:

> In order to maintain and extend their influence they [the leaders of parties representing newly-enfranchised workers] must command support from a mass following. Hence they will continue to oppose other elements of the ruling strata such as business and the aristocracy. The objective, however, of the mass-based elite is to replace the power of one [property-owning] minority with that of another, themselves.[65]

Among the contenders for the replacement ruling elite was Karl Marx with his authoritarian brand of communism. Michels, writing in 1910, couldn't know just how badly that would turn out.

Michels' 'Iron Law of Oligarchy' appears to be a special case of what might be called an 'Iron Law of Hierarchy'. Michels wanted to shed light on the divergence of the party from its voters. 'Oligarchy' is an appropriate word because it denotes a two-level structure – the party and the people. A hierarchy, in contrast, has multiple levels, and so does the internal structure of the political parties.

In 1867, when the franchise was widened in Britain, the Iron law of Oligarchy had not yet been recognised. The proponents of the wider franchise did not see how autocratic control could survive the reduction in power of the aristocracy with its various tiers of fox-hunting privilege.

But the new political leaders quickly worked out how they could preserve their power and patronage. Political parties began to demand and reward absolute loyalty from their elected representatives. Members stopped voting for their principles in the House of Commons. 'Party unity' took priority over ideas. It is not therefore paradoxical that the politician who legislated for the widening of the franchise – Disraeli – was a Conservative who up until then had served the previously dominant 'landed interest' in the Commons. Why should he worry that his widening of the franchise would strip his former patrons of their power? He could be confident that he could step in to fill the hierarchical vacuum and exercise the patronage himself. Disraeli would now be the organ-grinder instead of the monkey.

Two major parties would soon share power alternately between them, perpetuating the facade of "democratic" government. The parties together passed legislation to ensure that only their candidates could put their party affiliation on the ballot, and then compelled them to do so. Official parties can also put campaign messages on the ballot, but independent candidates cannot. Only the two largest partis can force a general election. They legislated to give parties public funding ('Short Money') related to how many votes they secure, which discourages tactical voting. These gradual constitutional changes have channelled power to the two main party hierarchies.

It is unlikely that any of the proponents of a wider franchise, except for power-mongers like Disraeli, would have wished for candidate selection by party elites, high expenditure on election campaigns, strict party loyalty rewarded by ministerial posts and honours and sinecures in the House of Lords, trained career politicians, and simplistic and divisive political ideologies.

Celebrities or Politicians?

James Madison in 1797 predicted that the personality cult could become another source of egregious party power:

> A zeal for different opinions concerning religion, concerning government...; an attachment ... to persons of other descriptions whose fortunes have been interesting to the human passions [read 'reality television'], have, in turn, divided mankind into parties, inflamed them with mutual animosity, and rendered them much more disposed to vex and oppress each other than to co-operate for their common good. So strong is this propensity of mankind to fall into mutual animosities, that where no substantial occasion presents itself, the most frivolous and fanciful distinctions have been sufficient to kindle their unfriendly passions and excite their most violent conflicts.

Central government takes back control

The devolution of fiscal power in Britain to local government in 1871 was a rare surrender of power by a central government which had no constitutional obligation to do so. Reaction was not slow in coming as the political oligopoly developed and took back the power it had lost.

The *coup* that gave final victory to the Iron Law of Oligopoly in Britain was the 1984 Rates Act's abolition of the local annual property tax that was financing the welfare state. Hardly anything needs to be said in favour of the UK's previous regular-revaluation annual property tax except that it has long been in use in all other wealthy countries.[66] All independent experts have advised successive British governments to return to it.

The stimulus for the abolition of this world-standard property tax forty years ago was that central government considered that it gave too much power to Local Authorities. Matters came to a head when elected councillors in Liverpool and Lambeth refused to obey the financial edicts of Whitehall. Kropotkin in his memoirs describes a similar situation in Tsarist Russia:

'Every attempt of the county councils (*Zémstvos*) to take the initiative in any improvement – schools, teachers' colleges, sanitary measures, agricultural improvement, etc – was met by the central government with suspicion, with hostility, – and denounced by the Moscow Gazette as "separatism," as the creation of "a state within the state," as rebellion …'

From 1990 successive British governments have prevented Local Authorities from levying proportionate value-based taxes on property, imposing a formula under which the lowest value property pays tax at five times the rate of the highest. This has made it impossible for local authorities to finance the welfare state.

As already explained, an annual property tax proportional to the up-to-date value of each individual property gives the local authority the financial ability to improve the public realm and public services. Improvements are self-financing by the increased tax yield from the properties that increase in value as a result. The two main British political parties have agreed to discontinue this almost world standard local tax in the interests of 'strong' (i.e. centralised hierarchical) government. They have increased centralised property taxes to compensate so that Britain pays as much property tax as anywhere. Every decision about local investment in public services must now be made by the ruling majority party in Whitehall. Lest it be thought that England is too small to warrant devolution of fiscal power, the US state of Rhode Island has a population and surface area comparable to Gloucestershire and has forty different municipalities which regularly revalue properties, set their own annual property tax rates, and plan their spending independently.

From the late 1980s, the centralisation of fiscal policy in Whitehall allowed the Conservative Party to begin its decades-long program of reducing public spending, based on a hypothesis that the welfare state was unfairly competing with private enterprise. The Labour Party, when it came to power, did not abandon its new absolute monopoly of control: it did not return any fiscal autonomy to local authorities.

The real advantage of local fiscal power

Admirers of localism – the setting and spending of taxes by independent local government – may think that its main advantage lies in local government's ability to understand local conditions.

From anarchism's perspective, however, something more important has been lost by Britain's abandonment of the world standard of local government. Local decision-making benefits from the innate ability of a community bound by human relationships to cooperate creatively even when the interests of individuals differ. What central government has done in Britain is to stamp out cooperation as a mechanism for administration and replace it by bureaucrats who rubber-stamp edicts from central government in 'branch offices of Whitehall' as Michael Heseltine, a former Conservative minister, described the hollowed-out remnants of local government.

The *coup* against local government was accompanied by the peremptory abolition of a complete tier of local democratic elected government – the Greater London Council – in 1986. Then in 1999 the majority party removed most hereditary, clerical, and judiciary members of the House of Lords, promising to convert it into an elected second chamber that would limit the power of the House of Commons. The promise was forgotten, and the second chamber is now dominated by titled unelected individuals who have given services or funds to the House of Commons political parties.

The concentration of power and dominance is possible in Britain because there is no enforceable written constitution to guarantee the autonomy of local elected bodies or to maintain checks and balances against concentration of power. Any power that local government had was always the gift of the majority party of the day.

The founders of democracy would have been horrified that the final achievement of the wider franchise in Britain is the evil most feared by the founding fathers: concentrated centralised power without checks and balances. Voters have achieved this by handing power to political party hierarchies.

Electoral politics: Hobson's choice

The optimum number of competing political parties, from hierarchy's point of view, is two. This rules out sharing power among a multitude of interests but satisfies the basic criterion of democracy. Each party chooses a flagship policy, enforced by party loyalty, that offers voters a choice. In Britain, the chosen alternatives are both too extreme for most voters. One party proposes expanding public ownership as much as possible, the other proposes to limit public ownership of public services to a minimum. Voters must therefore choose between voting against doctrinaire Marxism and voting against doctrinaire Anti-Marxism. No time is spent rationally discussing the factors that make an industry or service suitable for financing by risk capital, which is the principle advantage of privatisation. Voters, according to a report in 2017 from the Electoral Reform Society, vote for the party that they dislike the least.[67]

Winston Churchill thought that any version of democracy would be better than any alternative. Victorian political philosopher John Stuart Mill thought otherwise in his analysis of different forms of representative government.[68] For Mill, a certain type of democracy could be worse than an absolute monarchy:

> If, instead of struggling for the favours of the chief ruler [i.e. monarch], these selfish and sordid factions struggled for the chief place itself, they would certainly, as in Spanish America, keep the country in a state of ... despotism ... alternately exercised by a succession of political adventurers, and ... would have no effect but to prevent despotism from attaining the stability and security by which alone its [despotism's] evils can be mitigated or its few advantages realized.

'Political adventurers' seems to accurately describe the politicians who are driven by their and the media's obsession with strong leadership. John Stuart Mill's comparison to Spanish America suggests that if he could use modern English idiom he would describe Britain today as a banana republic.

CHAPTER 9

Roadmap to Democracy

S OME BRITISH POLITICIANS ARGUE THAT constitutional reform will make government more democratic. The Electoral Reform Society is in the forefront of the campaign. They argue that:

disillusionment, disengagement, and distrust are the words most often associated with people's relationship to representative politics … this is enabled by a voting system which hands one party undue power. Westminster's centralised system is holding this country back.

The Electoral Reform Society's strategy is to:

persuade one of the two major parties to back proportional representation. This is most likely to be the Labour party".[69]

Many countries use proportional representation, a voting system which results in smaller political parties winning more seats in the legislature. In Britain, one of the two major parties would have to agree to make this change by legislation, because they have between them around 90 per cent of the seats in Parliament. The smaller parties have only round 10 per cent of the seats, despite having between 20 or 30 per cent of the national vote. Their votes are so thinly spread that they cannot win a corresponding share of seats in Parliament.

The ERS is naïve in believing that it can persuade the Labour party to back proportional representation (PR), as history shows. In 1997 Labour was elected to power on the basis of its manifesto which promised a national referendum on proportional representation. It did not keep its promise. By the time of the next general election Labour merely promised to "review the experience of existing PR systems". It is not hard to see why a party that has tasted absolute power would backtrack on its promise to share it with smaller parties.

But there is something else wrong with the ERS strategy, apart from the fact that its plan for constitutional reform has no chance of success. Its goal (PR) is not a cure for what it identifies as people's main complaint, that is Westminster's centralised power. PR is simply the goal that obsesses the minor parties who are ERS's main supporters: they want more power in the democratic world's most dictatorial parliament. This shows that the risk of introducing PR is that the UK's total centralisation of power in the government of the day, unique in the democratic world, will corrupt smaller parliamentary parties as it has the main parties.

The corrupting influence on smaller parties was seen in action in 2010 when the Liberal Democratic Party won enough seats to stop either of the main parties forming a majority government alone. Instead of supporting either the minority government or opposition party on each issue according to their manifesto, the Liberal Democrats created a coalition with the biggest party and abandoned their manifesto pledges in order to gain experience in government (i.e. to join the oligarchy).

This chapter will explain how activists can peacefully undermine political parties and elect a real representative assembly using tactical voting. The examples quoted are from Britain, but the method is applicable to other countries. The opportunity in Britain is probably greatest because of the recent lamentable performance of the main political parties.

Power is not given, it is taken

Modern democracy did not result from peaceful evolution of earlier systems of government. It resulted from civil unrest which forced the wealthy to devise an electoral system which would allow the poor to participate but would still protect wealth. In the USA the impoverished troops who had just chased out the British monarchy were not prepared to submit to the rule of the wealthy. In Britain nearly a century later, one per cent of the national population arrived by railway at a forbidden rally for working class votes in London's Hyde Park. After demonstrating their capacity for violence they promised that the next rally would take place at the Houses of Parliament. The government, dominated by wealthy landowners, caved in and gave working men the vote.

The subsequent hijacking of power by political parties was not a spontaneous evolution either – it was a long drawn-out *coup*. It took a long time because the founders of democracy attempted to ensure that it could never happen by creating a fully representative assembly. They saw no advantage in replacing the tyranny of the rich by the tyranny of parties. They knew that partisanship would degrade the legislature. "Political parties," wrote John Marshall, Chief Justice of the United States from 1801 to 1835, "kindle the animosity of one part [of the nation] against another." The animosity that Marshall warned against is nowadays welcomed as the hallmark of democracy rather than its curse. Party theoreticians study long into the night to identify "wedge" issues, however trivial or irrelevant to people's welfare, that will divide the nation and demonstrate that the party system is 'necessary'. Our political parties are playing into the hands of the world's worst dictators who are gleefully mocking the social disharmony produced by these political parties in search of power.

The representative assembly that the founders devised to save their wealth from 'an equal division of property, or any other improper or wicked project', to quote James Madison, 4th President of the US, was not a preliminary version of today's political system but a failed attempt to prevent it.

DIY Electoral Reform

Now that nearly all adults have the right to vote, they can peacefully deprive the political parties of their illegitimate power. They can restore the representative assembly as it was conceived both by the first Athenian democrats and by their eighteenth and nineteenth century followers.

Already some political dissidents have begun using their votes in a way that could reform the legislative assembly. They have rejected the idea that the vote is simply a privileged opportunity for each citizen to express their private political beliefs. The dissidents instead see their vote as a way of combining with other members of the community to achieve immediate goals. This mode of voting is comparable to the procedure used in the world's first democratic legislative assembly, in ancient Athens. There, voting was open by show of hands, so people were influenced by the way their neighbours were voting. The same system is used in many 'town hall' type meetings in which repeated 'straw' (non-binding) votes are taken to see which way opinion is moving after speeches are made for and against a proposal.

I recall one 'town hall' meeting at which several hundred partners had to decide whether to merge with another firm. The head of one national office told me that if the merger took place he and his subordinates would resign and set up a rival firm. He therefore sat in the front row where he could be seen, and raised his hand against the merger proposal. The count showed that the partners were divided on the issue, so more speeches for and against were allowed. Finally, my dissident friend looked behind him before deciding which way to vote. He then abstained. Afterwards, I asked him why. "My people were not following me", he said. On this occasion, as on many others, it's not how *you* vote that is important; it's how *other people* vote.

In a similar way the electorate can use *tactical voting* in a general election to break the power of the political parties and restore a more representative assembly. One obstacle is the widespread misconception that the only meaningful vote is one that is cast for a political party. This misconception would have to be overcome so that the more democratic members of the electorate would adopt the new proposal that they vote *for whichever candidate was last in alphabetic order on the ballot paper*. This 'coordinated anti-party voting' would produce a number of MPs who, despite their nominal party affiliation, would have no loyalty to a party. It will not have been party affiliation that elected them, and it is unlikely that they will serve for more than one term or make a career in politics.

The electorate became familiar with tactical voting in the 2017 election. An activist group, Tactical2017, wanted voters to unite behind the progressive candidate, from whatever party, most likely to defeat the candidate of the Conservative Party. The Tactical2017 website listed all the national constituencies and showed which left-wing candidate was best placed to beat the Conservative, based on historical trends.[70] If the plan had worked, we would have replaced a right-wing dictatorship with a left-wing one.

The plan failed because a handful of constituencies failed to re-elect candidates from the biggest left-wing party after Labour: the Scottish Nationalists (SNP). Six 'safe' SNP seats switched to Conservative after the SNP alienated their voters by promising to hold a second referendum on Scottish independence.

The election results did demonstrate that many supporters of minority parties could be persuaded to 'hold their noses' and vote tactically for another party so that their votes could be more productive. If these voters had wanted local government to regain the responsibilities confiscated by the central government, they could have voted for the candidate most likely to beat both Labour and Conservative. Both of those parties when in government have failed to implement the European treaty that obliges them to empower local self-government, because they fear checks and balances against their arbitrary misuse of centralised power.

As another example of tactical voting, in the 2019 election some parties stood down in some constituencies to benefit the Unite to Remain [in the EU] campaign. Tactical voting is now understood by the electorate, but has aimed at promoting specific political parties or policies. The proposal put forward here has a different, more ambitious purpose: electoral reform without needing the permission of Parliament.

This proposal aims at limiting the accumulation of power and can therefore be seen as anarchist. 'Electoral anarchism' clearly does not involve civil disobedience.

How many voters would adopt the proposal?

We can use the results of the 2017 general election as a model. A large majority of 2017's voters (at least 80 per cent) would not have voted for the last candidate on the ballot in the way proposed here. There are, nevertheless, two groups of individuals who would have sympathised with the proposal and who could have disrupted the power of the political parties at the first attempt. They could have elected 100 'liberated' MPs from among the unsuccessful candidates – 100 rogue MP's who would have made it impossible for any realistic coalition of parties to form a government. These 100 MPs would represent their constituents, not the party which had selected them as a candidate.

The first group of potentially anti-party voters are diehard non-voters; the second group are those who voted for a 'no-hoper' party candidate.

In 2017, at least 20 per cent of the electorate refused to vote, even in the most hotly-contested constituency elections. We can safely assume that at least half of these (10 per cent) were dissatisfied with our political parties and would have voted to disrupt their power. Feeding this counterfactual assumption into the 2017 results shows that this 10 per cent of the electorate, alone, could have created 50 liberated MPs – half of the total needed for our DIY electoral reform

The second group – those who could be persuaded to vote for reform instead of for a no-hoper candidate – can also be quantified. 20 per cent of voters in 2017 chose a party that received less than half the votes received by the winner. Did they all know that their customary party was a no-hoper and that their vote would be wasted? I found when campaigning on the doorstep that most voters *did* know when they were on a loser, but they hadn't thought about how they could us their vote more constructively. When I told them that voting for me would ensure that my (more deserving) party would get more taxpayers' funds, they all expressed surprise. It seemed to me therefore that few voters are influenced by the 'Short Money' provisions that discourage tactical voting. So we can assume that anyone planning to vote for a party that will receive less than 50 per cent of the winner's votes can be persuaded to vote against the party system. The 2017-based model shows that this 20 percent of votes would create the other 50 liberated MPs needed to meet the target of 100.

It must be accepted that this revolution may proceed more slowly, possibly taking more than one human generation. Demographic changes will replace citizens who worship a political party by their children who will be aware of their bleak future prospects under this sham democracy that despotic nations are laughing at.

The simplicity of the proposed tactical voting algorithm will help to persuade voters. 'Last in alphabetical order' is not very mathematical, but voters will see immediately that it cannot be manipulated and is truly impartial.

The result of these anti-party votes would, in 2017, have created a representative assembly in which no realistic grouping of parties could have formed a stable coalition government. There would have been more than 100 different MPs elected simply because they were last on the ballot. They would have been converted into Independents. (In reality only one Independent was elected in 2017).

Electoral anarchism will reduce the power of the state

100 independent MPs free of any party discipline will resist any attempt by the two major parties to sideline them and dominate the chamber. The smaller parties (in this 'thought experiment' retaining about 50 of 650 seats) would combine with the 'independents' in this resistance. Cabinet ministers would have to be appointed by free parliamentary vote, not chosen by a party. The biggest party's ability to spend taxpayers' funds on grandiose projects with uncertain payback would be curtailed. These changes will erode the centralised power of the state and create the 'representative government' envisaged by democracy's American and British founders.

The news media (which now get free news from politicians in the dominant parties) would find new subjects to talk about. No longer would they focus on oversimplified 'wedge' issues designed to neatly divide the nation (privatisation, trade unions, immigrants, hunting, 'culture wars', gender identity).

One important new subject that will be openly discussed in parliament is the European Charter of Local Self-Government, an international treaty that Britain ratified in 1988 but which parliament refuses to implement. Implementation would increase economic growth by giving local government the power to levy proportionate property tax and spend it on short-payback infrastructure projects. This is what every other developed country does. Neither of the two major UK parties have made any move to implement the treaty when in power, and recently the Government took Scotland to court to prevent its implementation in that country alone. Although both dominant parties claim to be pursuing economic growth, it is clear that they will not sacrifice their joint monopoly of centralised power to achieve it.

Once the joint monopoly of power is eliminated, policies that pursue useful economic growth and prosperity will no longer be constrained by hierarchy's instinct for self-preservation.

Hierarchy tries to replace cooperation

Democratic local government benefits from the cooperative behaviour of councillors and their electors, neighbours who understand each other's points of view. This remarkable human behaviour appears to conflict with the laws of physics as embodied in the self-organising property of hierarchy. That self-organising property has enabled Britain's unchecked central government to neutralise cooperative democracy by depriving local government of fiscal power.

There are reasons why the state may choose to replace a cooperative venture with a hierarchical structure. The state decided to take over the postal system in order to intercept the private communications of suspected dissidents, as revealed by the Mazzini scandal of 1844. This surveillance extended into the telephone era, leading to state ownership there as well. Today, the state monopoly is not needed because the security services can tap into private telecommunications networks.

The foundation of the National Health Service (NHS) in Britain in 1948 offers another example of anti-cooperative state action. As Colin Ward points out in *Anarchism, A Very Short Introduction* (2004), the NHS was not a new system of health care. It was a state takeover of 'the vast network of friendly societies and mutual aid organisations that had sprung up through working-class self-help in the 19th century'. Herbert Morrison, the minister in charge of the nationalisation programme, favoured leaving these organisations in the care of municipalities, but was overruled by the Health Ministry.

The contributions that working class families had paid to their friendly society had not been taxed, but when the state took over, according to Ward, the revenue became 'the plaything of central government financial policy'.

It should be clear to the reader that spontaneous mutual aid delivers many if not all of the benefits that coercive government receives credit for. The examples above are of cooperatives that relied on thousands of individuals providing sometimes global services often using very advanced technology. If we ask ourselves whether mutual aid will expand faster than the state can during the next hundred years, we need to imagine a vastly different technological landscape.

The ability of humans to adapt to new technology at breathtaking speed may not be obvious to younger people who have not witnessed many changes. It is pointless trying to predict technological change, but there are developments that are already disrupting the state's domination of society. These include most obviously the internet, but also energy technology and super-powerful computing. These technological advances will continue to enable cooperation and undercut state power.

Non-hierarchical cooperation is already working in modern society, although our traditional worship of centralised power tends to obscure its success. Elsewhere I have shown how historians failed to record Victorian social reforms and humane military successes achieved in defiance of hierarchy. The importance of hierarchical control has been consistently inflated. The endless repetition of government successes and failures in the media have produced an almost universal delusion that nothing important happens without powerful hierarchical leadership.

But can anarchist cooperation build jumbo jets?

Some mega-products of our current society might appear to tax the capacity of cooperating human groups. Jumbo jets, nuclear power stations, and medical imaging and treatment technology: would we have to do without these and other boons to humanity, if the state were powerless? A federation of communities could put together a digital internet, but what about medical imaging and radiotherapy?

But as the scientist and anarchist Peter Kropotkin pointed out in 1899, already in his day there were complex technical cooperatives that did not need centralised hierarchies:

> [Groups can] combine directly, by means of free agreements between them, just as the railway companies or the postal departments of different countries cooperate now, without having a central railway or postal government ... or as the meteorologists, the Alpine clubs, the lifeboat stations in Great Britain ...'[71]

Nor must we ignore the fact that we would have had the World Wide Web at least two decades earlier if governments had not obstructed it by prohibiting the cooperative sharing of telecommunications circuits. Not only have we forgotten this state-imposed obstacle to cooperation, but there is a widespread false belief that only a state government could fund an internet, as the US Defense Department funded Arpanet. To quote from a recent study 'Only the state could afford the huge resources and the risks involved in launching the Internet'.[72] This endlessly repeated illusory truth seems to relate to the fallacy that warfare is the only stimulus for technological advance.

It is easy to refute the claim that only the state could fund the internet. The SITA airline reservation network exactly fitted Kropotkin's 1899 model of free agreements between groups. The network was a privately-funded civilian cooperative of the world's airlines. In 1972 it was already interconnecting different commercial enterprises using the well-understood public domain technologies later used by Arpanet including distributed computing, packet switching, and adaptive multipath routing. Unlike Arpanet it provided world-wide connectivity and high reliability, enabling travel agents and airline offices to make passenger reservations on multiple airlines simultaneously in real time. In the mid-1970s SITA began to upgrade its hardware and software to carry more real-time interactive traffic.

SITA's project succeeded on time and on budget while two similar projects failed, one of which was a dedicated procurement network funded by the US Department of Defense. The competing airlines coordinated SITA's work by sending delegates to meetings of the International Air Transport Association (IATA) in Montreal.

As demand for public access increased, The US Justice Department removed the prohibition on sharing circuits in 1982. Arpanet grew into an international internet cooperative. Only in the late 1990s did this public internet become reliable enough to take over the airline traffic from SITA. The internet only really took off at the beginning of the 21st century with 'web 2.0', when private individuals began to share information using communal websites funded by private sector investment, not by the state.

SITA could easily have provided the public internet. Its airline messaging cooperative had been functioning long before the arrival of computers. It had begun as a shared telegraph network, with banks of teletype machines connected to its satellite offices around the globe. In the switching centre in Paris, roller-skated attendants glided between machines with handfuls of punched paper tape for onward transmission. SITA's relationship with local circuit providers in the most remote parts of the globe, in conjunction with the emerging computer expertise of the cooperating airlines, made the transition to internet technology an easy task. While Arpanet researchers were still puzzling over how to unblock their network paralysed by a deluge of electronic Christmas greetings, SITA was ensuring that paying passengers were getting home to their families in time for the festivities.

Energy and democracy

The most obvious short-term technology that will undermine the hierarchy is non-fuel energy. Stored fuel resources — fossil, nuclear, or hydroelectric — have become a jealously-guarded asset of each state, to be disposed of to monopolies or used as a weapon against other states not so well endowed. New energy technology is destroying the strategic, as well as monetary, value of these stored fuels. Non-fuel energy (from wind, sunlight, and tides) can be a 'commons' exploitable by all without limit. This energy, unlike stored fuels such as oil, can no longer be amassed, hidden, diverted by corrupt politicians, speculated, monopolised, or used as a geopolitical instrument of power.

In Britain, the over-powerful state is slowing the economy's conversion to new energy technologies. Government ministers and the civil servants who report to them are directly participating in the energy market (their traditional role during the monopoly era) rather than adapting to the need to regulate the industry in the light of new competition-friendly technologies. The consequences of this hierarchical control include an energy network that is not 'joined up' and does not deliver electrical power to where it can be used, unambitious technology plans, and neglect of industrial diversification. The situation is analogous to the 'command economy' that existed before the agricultural revolution of the seventeenth century.

The network is not joined-up. The state itself is awarding contracts by auction to generators of wind and solar energy. They are proud of the fact that the winning bids show a rapid decline in cost with time. But the new generating equipment will be in locations far from the existing locations of older generating plant, and the transmission network upgrades needed are not included in the auction process. As more non-fuel generation is commissioned, the cost to the taxpayer of compensating generators for **not** generating the contracted power is rising fast.

Technology decisions are unambitious. The government's plan to 'digitalise' the electricity network, published in 2021, aims to use computers to gather operational data to facilitate central planning and reduce costs. These limited objectives compare poorly with the regulator's decision in 1990 to force the monopolistic telephone network to install computerised exchanges which reduced barriers to entry for competitors including mobile and internet providers. Britain's energy plan foresees that a user premises will still have only one supplier of electrical power. The goal of an 'energy internet' in which any generator can supply any consumer at uncontrolled prices is not mentioned.[73]

Industrial diversification is neglected. Britain's government has political preoccupations: reducing energy costs and national net greenhouse gas emissions are its top priorities. It has not sought to identify the potential of new energy technologies to change our way of life and industrial base, creating economic growth and prosperity. This is another example of failure to learn from the telecommunications revolution of 30 years ago. The omission also reflects the fact, mentioned earlier, that despite their public commitment to growth the dominant political parties will not sacrifice their oligarchy to achieve it.

We are the losers, because unexploited industrial uses for intermittent and interruptible electrical power will emerge, particularly when capacity becomes available that is surplus to predicted firm energy demand. Temporary surplus generating capacity is an inevitable result of intermittent generation destined to provide a fixed amount of firm power. Modelling shows that by 2035 there will be surplus UK wind generation capacity for about 50% of the time, and the surplus will exceed 25 GW for 10% of the year. By 2050 the surplus will be in the dozens of GW for half the year. Because this surplus results from seasonal weather patterns rather than from fluctuations in demand, it will often persist for days on end. Non-traditional electro-intensive industries such as indoor agriculture (using high-intensity LEDs to drive photosynthesis) will become viable. New manufacturing industries that can use intermittent production (including automated factories that produce only when the wind is blowing) will flourish. These activities may make it less necessary to rely on green hydrogen as an intermediate product; instead of storing power via an inefficient process it can be used to create value immediately. This is a strategy followed by Norway (aluminium production) and Chile (copper mining and smelting). The strategy is aligned with the intermittency patterns of those countries' zero-carbon energy resources. Interuptibility does not make a service valueless; it is still a feature of mobile communications, as it was for the wind and water power that began the Industrial Revolution.

One information processing application that will benefit from free off-grid energy is artificial intelligence (AI). Pessimists including the late Stephen Hawking have maintained that AI is a potential usurper of human intelligence. They may not be aware of the shortcomings of human intelligence resulting from natural selection's design of speech communication to suit a different environment. Our simplistic innate error detection protocol leaves us vulnerable to mass media onslaughts ranging from the daily newspapers to electronic propaganda robots. Technology has often served as a remedy for such limitations of natural selection, the most obvious example being clothing which allowed humans to colonise the higher latitudes. AI applications will detect information that is contaminated by bias, better than we can.

The carbon-free pedigree of intermittent capacity will also expand Britain's export opportunities. Britain, the EU, and other countries will have a carbon-pricing regime covering many of their domestic industries. A Carbon Border Adjustment Mechanism (import tariff) will be employed that puts imports on equal footing with domestically-produced goods of the same carbon footprint. Domestically-produced goods that have a *lower* carbon footprint will be at an advantage. This could address a common criticism of Britain's Net Zero goal: that it seeks to reduce *domestic* emissions only and that some of the reduction has been realised by abandoning manufacture and importing manufactured goods from countries that create more emissions than Britain did, thus aggravating global warming. An early decarbonisation of Britain's power sector holds out the prospect of 'reshoring' selected manufacturing capacity, for additional economic growth.

If a truly representative assembly were to remedy these defects in the UK's Net Zero project, the 50-year-old prediction of Britain's leading geographer, Professor Sir Dudley Stamp, would come true. He predicted that industry would relocate to countries where energy could be generated from local sources. Stamp's historical argument was that heavy industry first sprang up where the most valuable commodity — live energy — was cheapest. The weaving and metal fabrication industries in England grew up near the running water needed to drive the mills. As coal came into use to generate steam power, industry relocated to the coalfields. But when oil appeared, it was so cheap to transport that every country wanted to have its own heavy industry even if it had no local energy resources. The consequences, according to Professor Stamp, have been disruptive tariff barriers, trade disputes, and industrial inefficiency.

Britain's uniquely plentiful local energy resource is offshore wind. We have a maritime Exclusive Economic Zone which is among the largest in the world, where wind is plentiful and we have exclusive rights to exploit it. Our narrow island geography is ideally shaped to make use of it, with a high coastline-to-area ratio and no land more than 70 miles from the sea. Where it is necessary to use the wind energy at a distance from the nearest landing point, 'festoon' submarine power cables can transport it around the island with minimal need for unsightly new pylons. This is not a new technology; already Britain has international submarine power cable capacity equivalent to around one fifth of average electricity demand.

Right place, right time

A representative assembly not dominated by political parties would harvest plentiful low-hanging fruit by reversing oligopoly legislation that has accumulated over more than a century. The unchecked power that the political parties have usurped could be remorselessly turned against them in the House of Commons. Never would so many no-brainer reform opportunities have been presented to a legislature so empowered to take advantage of them. It would become apparent that the decline which has been blamed on debts run up in wars that Britain fought to defend others, or loss of Empire, was instead due to the main political parties' insatiable hunger for dictatorial power and patronage.

Simply restoring municipal government's tax and spend autonomy could save Britain's taxpayers £100bn annually. This is the roughly estimated opportunity cost of Britain's uniquely 'strong' central government which does not suffer local democracy and has therefore abolished annual property taxes based on regular revaluation. That kind of local property tax is the easiest tax to collect and to spend effectively in ways chosen by the taxpayers who paid it. It's a unique tax that makes the taxpayer richer. The rough estimate of saving comes from a comparison of tax burdens in the US and UK, allowing for differences in public spending in areas such as defence and health care.

Conclusions

It has been easy for the hierarchy to convince citizens of the need for hierarchical control because our weak evolved communication protocols make us susceptible to grooming by mass media. The state, through its education policies, exploits this weakness instead of correcting it.

Much of the daily communications we receive repeat the message that leadership is responsible for society's successes and failures. Frequent high-profile elections swell this steady flow of misinformation to a mighty torrent. The US political parties spent $6.5bn during the 2016 election campaign. This sum buys an irresistible chorus of election messages glorifying powerful leadership.

The prospect of an end to this obsessive pursuit of power over other people may seem remote. But Kropotkin believed that humanity's thought patterns would eventually change, as they changed before, because of the inevitable diffusion of new knowledge:

'Long years of propaganda and a long succession of partial acts of revolt against authority, as well as a complete revision of the teachings now derived from history, would be required before men would perceive that they had been mistaken in attributing to their rulers and their laws what was derived in reality from their own sociable feelings and habits.'[74]

Notes

[1] Harari, Yuval Noah. *Sapiens* Vintage, London 2014 p. 153, p. 115

[2] Quoted in Dugatkin, L., *The Altruism Equation: Seven Scientists Search for the Origins of Goodness.* Princeton University Press, 2006, p. 22

[3] Dugatkin, L., *op. cit.* p. 33

[4] Dugatkin, L., *op. cit.* p. 28

[5] Kropotkin, P., *Mutual Aid: A Factor in Evolution* ISBN 9781497333734 p. 8

[6] Kropotkin, *op. cit.* p. 45

[7] Maslin, M., Christensen, B., *Tectonics, orbital forcing, global climate change, and human evolution in Africa: introduction to the African paleoclimate special volume.*, Journal of Human Evolution 53 (2007)

[8] National Oceanographic and Atmospheric Administration, US Dept. of Commerce, http://www.ncdc.noaa.gov/paleo/ctl/abrupt.html viewed March 2012

[9] Richerson, P., and Boyd, R., *Not By Genes Alone: How Culture Transformed Human Evolution, Chicago:* University of Chicago Press, 2005 p. 257.

[10] Mameli, M., *Evolution and Psychology in Philosophical Perspective* in Barrett, L., and Dunbar, R., *Oxford Handbook of Evolutionary Psychology*

[11] Dunbar, R., Deacon's Dilemma: *The Problem of Pair-bonding in Human Evolution* in Dunbar, Gamble, and Gowlett (eds.) *Social Brain, Distributed Mind,* Proceedings of the British Academy vol 158, 2010

[12] Melis, A., et al. *Tolerance Allows Bonobos to Outperform Chimpanzees on a Cooperative Task,* Current Biology 17, 619–623, April 3, 2007. See also Barrett, L., and Dunbar, R., *Oxford Handbook of Evolutionary Psychology* p. 231 and the YouTube video at https://www.youtube.com/watch?v=zrv91Pa3jgs

[13] Maslin, M., *The Cradle of Humanity: How the challenging landscape of Africa made us so smart,* Oxford University Press, 2017 p.148

[14] Wrangham, R., *Catching Fire: How Cooking Made Us Human,* Basic Books, 2009, p.86

[15] Hawkes, K., *Showing Off,* Evolutionary Psychology Volume 6(2). 2008.

[16] Wrangham, R., *op. cit.* p. 135

[17] Dawkins, R., *The Selfish Gene* Oxford: Oxford University Press, 2006 p. 309

[18] Harari, Yuval Noah, *Op. cit.* p 75

[19] Dubreuil, B., *Human Evolution and The Origins of Hierarchies,* Cambridge: Cambridge University Press, 2010 p. 92

[20] Wrangham, R., *op. cit*, p. 185; Tomasello, M., *Why We Cooperate,* MIT Press, 2009 p. 84

[21] Thomas, N., *The Islanders: The Pacific in the Age of Empire*, London: Yale University Press, 2010

[22] Stringer, C., *The Evolution Of Our Species,* Penguin, 2012, p. 220

[23] Christenson, Morton H., *Language Evolution.* Oxford University Press 2003, p. 23

[24] Christensen, *op. cit.* p. 29

[25] It's easier to understand the mathematics of asking five people if one imagines five light bulbs that blink on and off at random, each being lit 80 per cent of the time. The question is then: at any given moment, what is the probability (i.e. proportion of time) that only 3, 4, or 5 lamps are lit. The formula is:

$$\sum\nolimits_{k=3}^{5} \binom{5}{k} p^k (1 - p)^{5-k}$$

This means "The number of combinations of exactly 3 lamps lit, multiplied by the probability of each combination, plus the number of combinations of exactly 4 lamps lit, multiplied by the probability of each combination, plus the number of combinations of all 5 lamps on (i.e.1), multiplied by the probability of that combination. (See Wikipedia on 'majority logic decoding' and 'Condorcet's Theorem')

[26] Christensen, *op. cit.* p. 37

[27] Dunbar, R., Annual Review of Anthropology Volume 32, 2003 pp 163-181

[28] Gamble, Gowlett, and Dunbar, *Thinking Big.* Thames & Hudson, 2014, p. 146; Dunbar, R. *The Social Brain: Mind, Language, and Society in Evolutionary Perspective,* Annu. Rev. Anthropol. 2003. 32:163-81

[29] Barker, G., *The Agricultural Revolution in Prehistory: Why Did Foragers Become Farmers?,* Oxford: Oxford University Press, 2006 p. 188

[30] Brody, H., *The Other Side of Eden: Hunter-Gatherers, Farmers, and the Shaping of the World.* Faber and Faber, 2002, pp. 148-9

[31] Lee, R., *The !Kung San: Men, Women and Work in a Foraging Society;* Cambridge University Press, 1979. Wrangham, R., *op. cit.* p. 143

[32] Brody, *op. cit.,* p. 14

[33] Arnold, B., and Wicker, N., *Gender and the Archaeology of Death,* Altamira Press, 2001 p. 4

[34] Barker, G., *op. cit.*

[35] Bloom, H., *The Book of J,* Faber, 1991

[36] Gray, J., *Heresies: Against Progress and Other Illusions.* London: Granta, 2004 p. 5

[37] Lovelock, J., *The Vanishing Face of Gaia: A Final Warning.* London: Allen Lane, 2009, p. 240

[38] Brody, *op. cit.,* p. 82

[39] http://haskelecon.blogspot.com/2013/02/can-intangible-investment-explain-uk.html retrieved 31 January 2019

[40] Lovelock, J., *op. cit.,* p. 53

[41] Maslin, *op. cit.* p. 168

[42] Description taken from Appleby, J., *op. cit.*

[43] Freeman, C., *A New History of Early Christianity*. New Haven: Yale University Press, 2009

[44] Laslett, P., *The World We Have Lost Further Explored*. London: Methuen 1983

[45] Google Ngrams (occurrences of a phrase in literature)

[46] Smith, A. *The Wealth of Nations*. Book II Ch. I: "To restrain private people ..."

[47] Michels, R. *A Sociological Study of the Oligarchical Tendencies of Modern Democracy*, London, Free Press, 1962.

[48] see Appleby, J. *The Relentless Revolution: A History of Capitalism*. New York: W. W. Norton, 2010. I have used Appleby's insightful account of the emergence of capitalism but I have also used and quoted her as a representative of the conventional (Weberian) wisdom.

[49] Appleby, J., *op. cit.* p. 87

[50] Seeley, Thomas D., Honeybee Democracy Princeton University Press, 2010

[51] Minto, B., *The Minto Pyramid Principle: Logic in Writing, Thinking and Problem Solving* (2010). Minto Books International Inc. London

[52] Narby, J., *The Cosmic Serpent*. London, Phoenix 1999

[53] Singer, E., *Brain Images Reveal the Secret to Higher IQ*, MIT Technology Review, March 24 2009

[54] http://geniusblog.davidshenk.com/2007/03/the_myelin_in_a.html

[55] Shenk, D., *The Genius in all of Us*, Icon Books 2010

[56] Jarvis Homes Ltd v Marshall & Anor [2004] EWCA Civ 839 (06 July 2004)

[57] Social Mobility and Child Poverty Commission, *Elitist Britain?* (August 2014)

[58] *Opportunities for your child in Birmingham 2017*. Birmingham City Council, Directorate for People School Admissions and Pupil Placements Service. https://www.birmingham.gov.uk/downloads/file/2629/secondary_education_o pportunities_in_birmingham_2017. Familiarisation brochure: https://www.birminghamgrammarschools.org/Birmingham_Familiarisation_Bo oklet.pdf

[59] *What prospects for mobility in the UK? A cross-national study of educational inequalities and their implications for future education and earnings mobility*. The Sutton Trust, 2011

[60] http://www.philosophyexperiments.com/wason/

[61] Cosmides, L. and Tooby, Adaptations for Reasoning About Social Exchange, in Buss, D. (ed.), Handbook of Evolutionary Psychology, Hoboken, New Jersey. Wiley, [2016]; Cheng, Patricia W., Holyoak, Keith J., Nisbett, Richard E., Oliver, L. M. (1986/07). *Pragmatic versus syntactic approaches to training deductive reasoning*. Cognitive Psychology 18(3): 293-328.

[62] Madison, J. *Federalist Papers, no. 10.*

[63] A vestige of political power remained with landlords because until 1872 voting was not secret so landlords could victimise tenants who voted incorrectly.

[64] See Jenkins, *Gladstone*, pp 287-317 and Matthew, *Gladstone 1809-1874*, pp 167-175
[65] Michels, *op. cit.* p. 18
[66] This land tax, in Britain, did not include a tax on 'undeveloped land' which is a more controversial tax often included under the rubric of Land Value Tax.
[67] Electoral Reform Society, *The 2017 General Election*. August 2017
[68] Mill, J.S. *Considerations on Representative Government*. Harper, New York 1862
[69] https://www.electoral-reform.org.uk/wp-content/uploads/2021/07/2021-2024-strategy-summary-1.pdf retrieved 31 March 2023
[70] https://www.tactical2017.com/#Gordon retrieved 19 January 2019
[71] Kropotkin, P. *Memoirs* p. 399 (Dover)
[72] Santos de Lima, L., Free Software Culture and Development. *Revista Crítica de Ciências Sociais, 6 | 2014*
[73] https://www.historyandpolicy.org/policy-papers/papers/digital-energy-a-history-lesson-from-telecoms retrieved 12 April 2023
[74] Kropotkin, P., *Memoirs* p. 400

Index

Author biography

Hugh Small graduated from Durham University with honours in physics and psychology. He became a business strategy consultant, helping governments and the private sector to modernise and liberalise the telecommunications industry.

Later he became a social historian, publishing extensively researched accounts of the social reforms and key political events of the Victorian era. His books expose conventional history's suppression of the achievements of anti-establishment reformers. The *Daily Telegraph* called his revolutionary biography of Florence Nightingale 'a masterly piece of historical detective work'.

His interests in both communications engineering and social reform developed in Chile while he was teaching Information Technology at the University of Concepción. The recently-installed military dictatorship of General Pinochet declared a State of Siege and martial law and prevented him from leaving the country for two years. He spent his spare time studying communications engineering journals in the university library and imagining the day when new network technology would liberate society.

After returning to Europe in the mid-1970s he spent the next six years helping to design and build the world's first commercial internet. The SITA network, which overtook the state-funded Arpanet, was a private sector collaborative venture between competing international airlines designed to allow travellers anywhere in the world to book seats on multiple carriers simultaneously.

His psychology training and early internet experience led him to the explanation in this book of how speech communication evolved in early human groups. He uses recent psychological studies to explain why our evolved speech mechanisms have 'vulnerabilities' in the very different modern environment.

In the 2015 general election he was the Green Party candidate for Cities of London and Westminster, receiving over 5 per cent of the vote.

www.ingramcontent.com/pod-product-compliance
Lightning Source LLC
Chambersburg PA
CBHW022044190326
41520CB00008B/702